Laxton's
TRADES PRICE BOOK

25

Laxton's
TRADES PRICE BOOK

Roofing 96-97

EDITED BY

Tweeds
CHARTERED QUANTITY SURVEYORS,
COST ENGINEERS, CONSTRUCTION ECONOMISTS

Published by Laxton's, an imprint of Butterworth-Heinemann
Linacre House, Jordan Hill, Oxford OX2 8DP

Edited by Tweeds, Chartered Quantity Surveyors, Costs Engineers and Construction Economists

© Tweeds, March 1996

ISBN 0 7506 2979 7

All rights reserved. No part of this book may be reprinted, or reproduced or utilised in any form or by any electronic, mechanical or other means, known now or hereafter invented, including photocopying and recording, or in any information storage and retrieval system, without permission in writing from the publisher.

The publisher and authors make no representation, express or implied, with regard to the accuracy of the information contained in this book and cannot accept any legal responsibility or liability for any errors or omissions that may be made.

A catalogue record for this book is available from the British Library

Printed and bound in Great Britain by Biddles Ltd, Kings Lynn and Guildford

Contents

Preface		vii
SMM6/SMM7 table		ix
Introduction		xi
1	Starting up in business	1
2	Running the business	17
3	Taxation	23
4	Estimating	35
5	Rates for measured work	41

RATES FOR MEASURED WORK

	Material costs	47
G	STRUCTURAL CARCASSING METAL/TIMBER	103
	G32 Woodwool slab decking	105
H	CLADDING/COVERING	109
	H30 Fibre cement sheet cladding	111
	H31 Metal profiled sheeting	112
	H41 Translucent sheeting	134

CONTENTS

	H60	Clay/concrete roof tiling	135
		Underfelt and battens	165
	H61	Fibre cement slating	167
	H62	Natural slating	173
	H63	Reconstructed stone slating	180
	H64	Timber shingling	182
	H71	Lead sheet coverings	183
	H72	Aluminium sheet coverings	191
	H73	Copper sheet coverings	194
	H74	Zinc sheet coverings	197
	H76	Fibre bitumen coverings	200
J		**WATERPROOFING**	205
	J41	Built-up felt roof coverings	207
L		**WINDOWS/DOORS/STAIRS**	225
	L11	Rooflights	227
Approximate estimating			257
6	Plant and tool hire		265
7	General data		269
Index			275

Preface

This book is written for the roofing contractor operating a small business or working as a sub-contractor.

Advice on the business side of the industry is contained in the early chapters and is set out in an easy-to-read style with many examples. The information given on current tax rates is taken from the Budget proposals of November 1994. It is not intended that these chapters should replace the need for professional advice when the occasion warrants it, but are meant to complement this need and hopefully save money in consultation fees.

Chapter 4 sets out some of the basic principles of estimating, but possibly the most valuable part of the book lies in Chapter 5 - Rates for Measured Work. The information contained in this chapter should help an estimator to produce quotations and tenders.

It cannot be stressed too strongly that despite claims made on their behalf, price books should not be used as a literal source of information for preparing quotations, but only as a base upon which contractors can prepare their own bids and tenders. The data provided are based upon costs prevailing in the third quarter of 1995.

The measured items are presented in accordance with the requirements of the seventh edition of the Standard Method of Measurement of Building Works (SMM7). Although this may be irrelevant to the needs of smaller contractors it would be required by firms bidding for work on tender documents based upon SMM7.

The indexing and reference systems are prepared in a style that the self-employed contractor who prices jobs without formal tender documents can also receive the maximum benefit from the book. It should be noted, however, that not all of the requirements of SMM7 have been observed and it is hoped that the compromise will produce a balance which will suit the majority of readers.

There are many items affecting the value of roofing works. The speed and skill of individual workmen and the price paid for materials are factors which can directly affect the profitability of a job and the rates in Chapter 5 should be used only as a basis for preparing an estimate or quotation.

PREFACE

These comments do not detract from the real value of the information in this chapter. Over 4,000 item descriptions and unit rates are given which are based on 15,000 separate pieces of data providing a wealth of detailed cost information.

A great deal of help has been received many sources in the research and preparation of this book and I would like to acknowledge the assistance by the following organiazations and individuals with grateful thanks:

Redland Roof Tiles Ltd	Kirkstone Green Slate Quarries Ltd
Ruberoid Building Products	Eternit TAC Ltd
Marley Roof Tiles Ltd	The Velux Company Ltd
Coxdome Ltd	Colt Building Products
Europroof Ltd	Pittsburgh Quarries Ltd
Precision Metal Forming Ltd	Penryhn Quarries Ltd
British Nuralite PLC	Aerial Plastics Ltd
Torvale Building Products Ltd	Jewsons
Merseyside Slate and Tile Company	HSS Hire Shops
Cookson Industrial Materials Ltd	Paul Spain

Grant Thornton provided the advice on business and taxation matters set out in Chapters 1 to 3. I would particularly like to thank Bryan Spain who is responsible for the original concept and compilation of the book's contents.

I would welcome constructive criticism of the book together with suggestions for improving its scope and contents. Whilst every effort has been made to ensure the accuracy of the information given in this publication, neither Tweeds nor the publishers in any way accept liability of any kind resulting from the use made by any person of such information.

The male pronoun is used in this book. This is for ease of reading and should be taken to mean both male and female individuals.

Christopher Powell
Managing Partner
TWEEDS
Chartered Quantity Surveyors
Cavern Walks
8 Mathew Street
Liverpool L2 6RE

SMM6/SMM7 Table

The following table is intended to assist the reader identify the new work sections in SMM7 by listing them adjacent to the same elements in SMM6.

	SMM6	SMM7
M	ROOFING	H CLADDING/COVERING
	Slate or tile roofing	H60 Clay/concrete roof tiling H61 Fibre cement slating H62 Natural slating H63 Reconstructed stone slating/tiling H64 Timber shingling H76 Fibre bitumen thermoplastic sheet coverings/flashings
	Corrugated or troughed sheet roofing or cladding	H30 Fibre cement profiled sheet cladding covering/siding H31 Metal profiled/flat sheet cladding/covering/siding H32 Plastics profiled sheet cladding/covering/siding H33 Bitumen and fibre profiled sheet cladding/covering H41 Glass reinforced plastics cladding/features K12 Under purlin/inside rail panel linings G30 Metal profiled sheet decking

SMM6/SMM7 TABLE

SMM6	SMM7
M ROOFING (cont'd)	
Roof decking	G31 Prefabricated timber unit decking
	G32 Edge supported/reinforced woodwool slab decking
	J22 Proprietary roof decking with asphalt finish
	J43 Proprietary roof decking with felt finish
	K11 Rigid sheetflooring/sheathing/linings/casings
Bitumen-felt roofing	J41 Built-up felt roof coverings
	J42 Single layer plastics roof coverings
Sheet metal roofing	H70 Malleable metal sheet prebonded coverings/cladding
Sheet metal flashings and gutters	
	H71 Lead sheet coverings/flashings
	H72 Aluminium sheet coverings/flashings
	H73 Copper sheet coverings/flashings
	H74 Zinc sheet coverings/flashings
	H75 Stainless steel sheet coverings/flashings
Protection	A42 Contractor's services and facilities

Introduction

There are two main ways that small firms and sub-contractors prepare tenders and quotations. The first and probably the less frequent method is to insert rates against item descriptions in a tender document prepared by the client's professional advisers. This would occur on major schemes where a bill of quantities and/or a schedule of rates has been prepared. In this case the contractor would be able to examine and use the rates contained in Chapter 5 of this book as a basis for his bid.

Usually, however, the contractor is either handed a plan and specification or merely invited to inspect the premises and prepare his offer without the benefit of any paperwork at all! In both of these two cases the contractor must take off his own quantities, but once he has done that the rates in Chapter 5 can be used.

Whichever method is used, the main value of this book lies in a sensible application of the thousands of rates in Chapter 5. People who have not used price books before sometimes query their value because craftsmen do not work at the same speed, wide variations in discounts for materials can be obtained, each job has different cost-related circumstances, etc.

The answer to these criticisms is that successful users of price books are fully aware of these difficulties but overcome them by understanding the relationship between their own production costs and material discounts and those assumed in the book.

For example, in using the book regularly, it may be found that the published rates are higher or lower from established costs. Quotations can still be prepared quickly by using the book rates and making the appropriate percentage adjustment at the end.

Careful thought must be given to the unique circumstances of each job undertaken. Profits can turn into losses without an assessment of the effect that each project's unique features can have on the rates, and the editors urge readers to consider this matter very carefully.

A superficial view of construction estimating believes that it is a precise exercise but there are too many variables involved for this to be true and the best any estimator can hope for is to foresee possible risks and make allowances for them. It is hoped that this book can help in this process.

Chapter 1
Starting Up in Business

There has been little improvement in the economy in recent years with hardly any evidence of economic growth. On the contrary, many new markets have reduced and existing businesses along with the majority of new start-ups have continued to fail.

It is increasingly more important that before giving up his job, the would-be businessman should consider carefully whether he has the required skills and the temperament to survive in the highly competitive self-employed market.

Before commencing in business it is essential that he makes a self-assessment on the commercial viability of his intended business because no matter how obvious this may sound there is no point in struggling to finance a business which is not commercially viable. In the early stages it is important to make decisions such as: What exactly is the product being sold? What is the market view of that product? What steps are required before the developed product is first sold and where are those sales coming from?

Research and seek out as much information as possible about how to run a business. Production schedules should be set and it should be clearly established what is required in order that those important first sales are obtained. Finally, do not underestimate the amount of time required to establish and finance a new business venture.

Whatever the size of the business it is important that you put in writing exactly what you are trying to do. This commonly takes the form of a business plan which can have a dual function of assisting you in establishing and monitoring your business and also providing an essential tool in your ability to raise finance. The contents of a typical business plan are discussed later in this chapter. It is important to realise that you are not always on your own and there are many contacts and advising agencies which can be of assistance.

STARTING UP IN BUSINESS

INITIAL RESEARCH

Potential customers and trade contacts

Many persons intending to commence in business on their own account in the construction industry already have experience as employees. Use these contacts to check the market, establish the sort of work which is available and the current contract rates.

In the domestic market, check on the competition for prices, standards of work and service provided, customer complaints and types of advertising. Try to get firm promises of work before the start-up date.

Testing the market

Talk to as many traders as possible who are operating in the same field. Try to identify if the need is in the industrial, commercial, local government or domestic field. Talk to likely customers and clients and consider whether it is possible to improve on what they are being offered in terms of price, quality, speed, convenience and follow-up.

INITIAL INFORMATION

Business Links

There is no shortage of information about the many aspects of starting and running your own business. Finance, marketing, legal requirements, developing your business idea, and taxation are all the subject of a mountain of books, pamphlets, guides and courses, and it should not be necessary to pay out a lot of money for this information. Indeed, the likelihood is that the aspiring businessman will be overloaded and thoroughly confused rather than left high and dry without guidance.

Business Links are now well established and provide a good place to start for both information and advice. There are organisations established with a view to providing a 'one-stop-shop' for advice and assistance to owner-managed businesses.

They will often replace the need to contact Training and Enterprise Councils (TECs) and many of the other official organisations listed below.

Point of contact: telephone directory for address.

INITIAL INFORMATION

Training and Enterprise Councils (TECs)

TECs are comprised of a board of directors drawn from the top men in local industry, commerce, education, trade unions etc, who together with their staff and experienced business counsellors, assist both new and established concerns in all aspects of running a business. This takes the form not only of across-the-table advice but also, if desired, hands on assistance in management, marketing and finance. There are also training courses and seminars available in most areas.

Point of contact: local Jobcentre or Citizens' Advice Bureau.

Enterprise agencies

Another useful source of free help and advice is the local Enterprise Agency. Run by local businesses for small and developing concerns it covers similar ground to the Business Links.

Point of contact: local telephone directory.

Banks

Approach banks for information about the business accounts and financial services that are available. Your local Business Link can advise on how best to find a suitable bank manager and to inform you on what he will require.
 Shop around several banks and branches if you are not satisfied at first. Managers vary widely in their views on what is a viable business proposition. Remember, most banks have useful free information packs to help business start-up.

Point of contact: local bank manager.

HM Inspector of Taxes

Make a preliminary visit to the local tax office enquiry counter for their publications:

IR 14/15	*Construction Industry Tax Deduction Scheme*
IR 24	*Class 4 National Insurance Contributions*
IR 28	*Starting in Business,*
	and, if needed,
IR 40	*Conditions for Getting a Sub-Contractor's Tax Certificate*
IR 53	*PAYE for Employers (if you employ someone)*

STARTING UP IN BUSINESS

IR 56/N139   *Employed or Self-Employed*
IR 57   *Thinking of working for yourself?*
IR 104   *Simple Tax Accounts*
IR 105   *How Your Profits are Taxed*

Remember the onus is on the tax payer to notify the Inland Revenue that he is in business. Failing to do so may result in the imposition of interest and penalties. Either send a letter or use the form provided in the middle of the *'Starting in Business'* booklet.

Point of contact: telephone directory for address.

VAT

The VAT office also offer a number of useful publications including;

700	*The VAT Guide*
700/1	*Should I be Registered for VAT?*
700/12	*Filling in Your VAT Return*
700/21	*Keeping Records and Accounts*
708	*Buildings and Construction*
731	*Cash Accounting*
732	*Annual Accounting*
742	*Land and Property*

Information about the Cash Accounting Scheme and the introduction of annual VAT returns are contained in Chapter 3.

Point of contact: telephone directory for address.

DSS office

Ask at the DSS office for the following publications:

NI 41   *NI Guide for the Self-Employed*
NI 27A/CA02 *People with Small Earnings from Self-Employment*
NI 35/CA44 *NI for Company Directors*
NI 39   *Employed or Self-Employed*
NI 225/CA04 *Direct Debit - The Easy Way to Pay,*

and for Employers

INITIAL INFORMATION

NP 18/CA03 *Class 2 and Class 4 NI Contributions*
NP 15 *Employer's Guide to NIC*
NP 227 *Employer's Guide to Statutory Sick Pay*

Point of contact: telephone directory for address

Local authorities

Authorities vary in provisions made for small businesses but all have been asked to simplify and cut delays in planning applications. In Assisted Areas and Enterprise Zones rent-free periods and reductions in rates may be available on certain industrial and commercial properties. As a preliminary to either purchasing or renting business premises, the following booklets will be helpful:

A   *Step by Step Guide to Planning Permission for Small Businesses* and *Business Leases and Security of Tenure*

Both are issued by the Department of Employment and are available at council offices, Citizens' Advice Bureaux and TEC offices. Some authorities run training schemes in conjunction with local industry and educational establishments.

Point of contact: usually the planning department - ask for the Industrial Development or Economic Development Officer.

Department of Trade and Industry

The services formally provided by the Department are now increasingly being provided by Business Link . The Department can still, however, provide useful information on available grants such as Regional Selective Assistance (RSA) and Regional Enterprise Grant (REG).

Point of contact: telephone 0171-215 5000 and ask for the address and telephone number of the nearest DTI office and copies of their explanatory booklets.

Department of the Environment

From 1 April 1992 new regulations came into force are in force under the Environmental Protection Act 1990 relating to all forms of waste other than normal household rubbish. Any concern which produces, stores, treats, processes, transports, recycles or disposes of such waste has a 'duty of care' to ensure it is properly discarded and dealt with. Practical guidance on how to comply with

STARTING UP IN BUSINESS

the law (it is a criminal offence punishable by a fine not to) is contained in a booklet *Waste Management: The Duty of Care: A Code of Practice,* obtainable from HMSO Publication Centre, PO Box 276, London SW8 5DT. Telephone 0171-873 9090. Price £5.25.

Accountant

The services of an accountant are to be strongly recommended from the beginning because the legal and taxation requirements start immediately and must be properly complied with if trouble is to be avoided later. A qualified accountant must be used if a limited company is being formed and for all types of business the accountant should be able to give advice on a whole range of business issues including book-keeping, tax planning and compliance to finance raising and will help in preparing annual accounts.

It is worth spending some time finding an accountant who has other clients in the same line of business and is able to give sound advice particularly on taxation and business finance and is not so overworked that damaging delays in producing accounts are likely to arise. Ask other traders whether they can recommend their own accountant. Visit more than one firm of accountants, ask about the fees they charge and how much the production of annual accounts and agreement with the Inland Revenue are likely to cost and how long the work will take. A good accountant is worth every penny of his fees but do not hesitate to challenge him if his service is unsatisfactory.

Solicitor

Many businesses operate without the services of a solicitor but there are a number of occasions when legal advice should be sought. In particular no-one should sign a lease of premises without asking a solicitor what they are committing themselves to because it is not unusual for a business to be put into financial difficulty through unnoticed liabilities in its lease. Either an accountant or solicitor will help with drawing up a partnership agreement which all partnerships should have.

A solicitor will also help to explain complex contractual terms and prepare draft contracts if the type of business being entered into requires them.

Insurance broker

Policies are available to cover many aspects of business including:

> employer's liability - compulsory if the business has employees
> public liability - essential in the construction industry

INITIAL INFORMATION

motor vehicles
theft of stock, plant and money
fire and storm damage
personal accident and loss of profits.

Brokers are independent advisers who will obtain competitive quotations on your behalf. See more than one broker before making a decision - their advice is normally given free and without obligation.

Point of contact: telephone directory or write for a list of local members to:

The British Insurance Brokers Association
Consumer Relations Department
BIBA House
14 Bevis Marks
London
EC3A 7NT (telephone: 0171-623 9043)

or contact

The Association of British Insurers
51 Gresham Street
London
EC2V 7HQ (telephone: 0171-600 3333)

who will supply free a package of very useful advice files specially designed for the small business.

The Health and Safety Executive

The Executive operates the legislation covering everyone engaged in work activities and the following free literature is available:

HSE 16 *The Law on Health and Safety at Work*
IND(G)14(L) *Compliance with Health and Safety Legislation at Work*
HSE4 *Short Guide to Employer's Liability (Compulsory Insurance) Act 1969*

The Executive has issued a very useful set of 'Construction Health Hazard Information Sheets' covering such topics as handling cement, lead and solvents, safety in the use of ladders, scaffolding, hoists, cranes, flammable liquids,

STARTING UP IN BUSINESS

asbestos, roofs and compressed gases. etc. A pack of these may be obtained free from your local HSE office or The Health & Safety Executive Central Office, Sheffield (telephone: 01142-892345) or HSE Publications (telephone: 01787-881165).

BUSINESS PLAN

As stated above, once available information has been obtained it is then recommended that this is consolidated into a formal business plan. The complexity of the plan will depend in the main on the size and nature of the business concerned. Consideration should be given to the following points.

Objectives

It is important to establish what you are trying to achieve both for you and the business. A provider of finance may be particularly influenced by your ability to achieve short- and medium-term goals set and have confidence in continuing to provide finance for the business. From an individual point of view it is important to establish goals because there is little point in having a business which only serves to achieve the expectations of others whilst not rewarding the would-be businessman.

History

If you already own an existing business then commentary on its existing background structure and history to date can be of assistance. There is no substitute for experience and any existing contacts which you have in the construction industry will be of assistance to you.

The following points should also be considered for inclusion:

- a brief history of the business identifying useful contacts made

- the development of the business, highlighting significant successes and their relevance to the future

- principal reasons for taking the decision to pursue this new venture

- details of present financing of the business.

INITIAL INFORMATION

Products or services

It is important to establish precisely what it is you are going to sell. Does your product have any unique qualities which gives it advantages over those of competitors? For example, do you have an ability to react more quickly than they do and are you perceived to deliver a higher quality product or service? A business plan would typically include:

- description of the main products and services

- statement of disadvantages whilst advising how they will be overcome

- details of new products and estimated dates of introduction

- profitability of each product

- details of research and development projects

- after-sales support.

Markets and marketing strategy

This section of the business plan should show that thought has been given to the potential of the product. In this regard it can often be useful to identify major competitors and make an overall assessment of their strengths and weaknesses and should also include the following:

- an overall assessment of the market, setting out its size and growth potential

- a statement showing your position within the market

- an identification of main customers and how they compare

- details of typical orders and buying habits

- pricing strategy

- anticipated effect on demand of pricing

- expectation of price movement

- details of promotions and advertising campaigns.

STARTING UP IN BUSINESS

It is important to identify who your customers are and why they might buy from you. Those entering the domestic side of the business will need to think about the best way to reach potential customers. Are local word-of-mouth recommendations enough to provide reasonable work continuity. If not, what is the most effective method of advertising to reach your customer base?

Remember, advertising is costly. It is a waste of funds to place an advertisement in a paper which circulates in areas A, B, C & D if the business can only cover area A.

Research and development

If you are developing a product or a particular service, then an assessment should be made on what stage it is at and what further finance is required to complete it. It may also be useful to make an assessment on the vulnerability of the product or service to innovations being initiated by others.

Basis of operation

Detail what facilities you will require in order to carry on your trade in the form of property, working and storage areas, office space, etc. An assessment should also be made on the assistance you will require from others. Your business plan might include:

- a layman's guide to the process or work

- details of facilities, buildings and plant

- key factors affecting production such as yields and wastage

- raw material demand and usage.

Management

This section is one of the most important because it demonstrates the capability of the would-be businessman. The skills you need will cover production, marketing, finance and administration. In the early stages these can be provided by the would-be businessman. However, as the business grows then it may be required to develop a team with individual attributes to cover each function. A full range of qualities are rarely present to a significant degree in one individual. The following points should be considered for inclusion:

INITIAL INFORMATION

- set out age, experience and achievements

- state additional management requirements in the future and how they are to be met

- identify current weaknesses and how they will be overcome

- state remuneration packages and profit expectations

- give detailed CV's in appendices.

Advertising and retraining may be required in order to identify and provide suitable personnel where expertise and experience are lacking.

Financial information

It is important to detail, if any, the present financial position of your business and the budgeted profit and loss accounts, cash flows and balance sheets. These integrated forecasts should be prepared for the next twelve months at monthly intervals and annually for the ensuing two years.

If the forecasts are to be reasonably accurate then the businessman must make some early decisions about:

- premises where the business will be based, the initial repairs and alterations that might be required and an assessment of the total cost

- what plant, equipment and transport is needed, whether it is to be leased or purchased and what the cost will be?

- how much stock of materials, if any, should be carried - the bare minimum only should be acquired so a reliable supplier should be sought out

- what will be the weekly bills for overheads, wages and the proprietor's living costs?

- what type of work is going to be undertaken, how much profit margin can realistically be obtained and how often are invoices to be presented?

STARTING UP IN BUSINESS

Your business plan should include:

- explanations of how sales forecasts are prepared
- levels of production
- details of major variable overheads and estimates
- assumptions in cash flow forecasts e.g. credit taken and given
- details of any tax assumptions (VAT/PAYE/Income Tax/Corporation tax)
- inflation assumptions.

Finance required and its application

The detailed financial information above will enable an accurate assessment of the funds required in order to finance the business. It is important to distinguish between those items which require permanent finance and those which ultimately will be converted to cash. This is because it is usually advisable to finance permanent long term assets with personal equity or loans repayable over similar periods.

Working capital such as stock and debtors can usually be financed by a more flexible loan arrangement such as an overdraft. Your bank manager can be of assistance in discussing the types of finance available.

Executive summary

Although it is prepared last, this will usually be the first part of your business plan. Remember business plans are prepared for busy people and their judgement may be based solely on this section. It will cover no more than two or three pages and will deal with all the important aspects and opportunities available within your plan. The summary should cover:

- the purpose of writing the business plan
- the major elements of the plan, including
    - key strategies
    - finance required and how it is to be used

BUSINESS PLAN

- management experience and its relevance

- expected returns showing potential rewards

- markets.

- appendices

- CV of key personnel

- organization charts

- market studies

- product advertising literature

- professional references

- financial forecast

- glossary of terms

To the extent that you feel any detailed information should be provided in support of your argument, then this is usually best provided in the appendices.

Follow up

Please remember once your plan is prepared, it is important constantly to re-examine it and update the forecasts and financial information. This is a working document and can be an important tool in running the business.

SOURCES OF FINANCE

Personal funds

Finance, like charity, often begins at home and a would-be businessman should make a realistic assessment of his net worth, including the value of his house after deducting the mortgage(s) outstanding on it, savings, any car or van owned and any sums which his family are prepared to contribute but deducting any private borrowings which will come due for payment. The whole of these funds may not be available (for instance, money which has been loaned to a friend or relative who is known to be unable to repay at the present time).

STARTING UP IN BUSINESS

It may not be desirable that all capital should all be put at risk on a business venture so the following should be established:

- how much cash you propose to invest in the business

- whether the family home will be made available for any business borrowing

- state total finance required

- how finance is anticipated being raised

- interest and security to be provided

- expected return on investment.

Whilst it may be wise not to pledge too much of the family assets, it has to be remembered that the bank will be looking closely at the degree to which the proprietor has committed himself to the venture and will not be impressed by an application for a loan where the applicant is prepared to risk only a small fraction of his own resources.

Having decided how much of his own funds to contribute, the businessman can now see the level of shortfall and consider how best to fill it. Consideration should be given to partners where the shortfall is large and particularly when there is a need for heavy investment in fixed assets such as premises and capital equipment. It may be worthwhile starting a limited company with others also subscribing capital and to allow the banks to take security against the book debts.

Banks

The first outside source of money to which most businessmen turn is the bank and a book could be written solely on the 'do's & don'ts' of approaching a bank manager. Here are a few tips:

- present your business plan to him. Please remember to use conservative estimates which tend to understate rather than overstate the forecast sales and profits

- know the figures in detail and do not leave it to your accountant to explain them for you. The bank manager is interested in the businessman and not his advisers and will be impressed if the businessman demonstrates a certain grasp of the financing of his business

- understand the difference between short and long term borrowing

SOURCES OF FINANCE

- ask about the Government Loan Guarantee Scheme if there is a shortage of security for loans. Under this scheme the Government guarantees 70% of the loan up to £100,000 (now up to 85% for loans to businesses in the inner-city taskforce areas, but if you have been in business for more than two years the limit is £250,000. In return there is an insurance of 0.5% on fixed-rate interest arrangements and 1.5% on variable bank interest arrangements. Repayment is over a period from two to seven years.

Remember the bank will want their money back, so bank borrowings are usually required to be secured by charges on business assets. In start-up situations, personal guarantees from the proprietors are normally required. Please ensure if these are given that they are regularly reviewed to see if they are still required.

Enterprise Investment Scheme - business angels

If an outside investor is sought in a business he will probably wish to invest within the terms of the Enterprise Investment Scheme which enables him to gain income tax relief at 20% on the amount of his investment. Additionally, any investment can be used to defer capital gains tax. The rules are complex and professional advice should always be sought.

Hire purchase/leasing

It is not always necessary to purchase outright assets required for the business and leasing and hire purchase can often form an integral part of a business's medium-term finance strategy.

Venture Capital

In addition, there are a number of other financial institutions in the venture capital market that can help well established businesses, usually limited companies, who wish to expand. They may also assist well-conceived start-ups. They will provide a flexible package of equity and loan capital but only for large amounts, usually sums in excess of £150,000 and often £250,000.

Usually the deal involves the financial institution having a minority interest in the voting share capital and a seat on the board of the company. Arrangements for the eventual purchase of the shares held by the finance company by the private shareholders are also normally incorporated in the scheme.

STARTING UP IN BUSINESS

The Royal Jubilee and Princes Trust

These Trusts through the Youth Business Initiative provide bursaries of not more than £1,000 per individual to selected applicants who are unemployed and age 25 or over. Grants may be used for tools and equipment, transport, fees, insurance, instruction and training but not for working capital, rent and rates, new materials or stock. They operate through a local representative whose name and address may be ascertained by contacting the Prince's Youth Business.

Point of contact: telephone 0171-321 6500.

The Business Start-up Scheme

This is an allowance of £40 per week, in addition to any income made from your business, paid for forty weeks.

To qualify you must be at least 18 and under 65, work at least 36 hours per week in the business and have been unemployed for at least six weeks or fall into one of the other categories - disabled, ex-HMS or redundant.

The first step is to get the booklet on the subject from your local Jobcentre or TEC, which includes details on how and where to apply. Once in receipt of the enterprise allowance, you will also have the benefit of advice and assistance from an experienced businessman from your TEC. All the initial counselling services and training courses are free.

Chapter 2
Running the Business

Many businesses are run without adequate information being available to check trends in their vital areas, e.g. marketing, money and managerial efficiency.
It is vital to look critically at all aspects of the business to maximise profits and eliminate inefficiency. Regular meaningful information is required on which management can concentrate. This will vary according to the proprietors' business but will often concentrate on debtors, creditors, cash, sales and orders.

Proprietors often have the feeling that the business should be 'doing better' than it is, without being able to identify what is going wrong. Sometimes there is the worrying phenomenon of a steadily increasing work programme coupled with a persistently reducing bank balance or rising overdraft.

Some useful ways of checking the position and identifying problem areas are given below.

Marketing

Whilst management and finance are concerned with the internal running of a business, the market is where it makes contact with the outside world in the shape of its competitors. Throughout his business life the entrepreneur should study carefully the methods and approach of the former and the needs and wishes of the latter. A shortcoming frequently found in ailing concerns is that the proprietor thinks he knows better than his customers what they want.

The term 'market research' sounds both difficult and expensive but a very simple form of it can be done quite effectively by the businessman and his sales staff. The type of person or business to whom the products or services are likely to appeal should be indentified, and then finding out and recording what the customer wants in terms of price, quality, design, payment terms, follow-up service, guarantees and services.

The initial approach might be by leaflet or letter followed by a personal call.
As an on-going part of management all staff with customer contact should be encouraged to enquire about and record customer preferences, complaints, etc. and feed it back to management.

RUNNING THE BUSINESS

Other sources of information can be trade and business journals, trade exhibitions, suppliers and representatives from which information about trends, new techniques and products can be obtained and studied.

Valuable information can also be gained from studying competitors and the following questions should be asked:

- what do they sell and at what prices?

- what inducements do they offer to their customers - e.g. credit facilities, guarantees, free offers and discounts?

- how do they reach their customers - local/national advertising, mail shots, salesmen, local radio and TV?

- what are the strongest aspects of their appeal to customers and have they any weaknesses?

The businessman should apply all the information gathered from customers and competitors to his own range of products with a view to making sure he is offering the right product at the right price in the most attractive way and in the most receptive market.

In a small business where the proprietor is also his own salesman he must give careful thought on how he can best present his product and himself. For instance, if he is working solely within the construction industry his main problems are likely to centre on getting a 714 Certificate and exploiting as fully as possible trade contacts to get sub-contract work.

However, for those who serve the general public, presentation can be a vital element in getting work. The customer is looking for efficiency, reliability and honesty in a trader and quality, price and style in the product. To bring out these facets in discussion with a potential customer is a skilled task. A short course on marketing techniques could pay handsome dividends. The Business Link will give the names and addresses of such courses locally.

Financial control

Unfortunately some firms which close down did not seek financial advice until it was too late to halt the downward trend. Earlier attention to the problems may have saved some of them. There are many reasons for this and one of them is that those running the business are unable to recognise the tell-tale signs and very few accountants take the trouble to explain to their clients what to look for. There are some tests and checks that can be done quite easily.

FINANCIAL CONTROL

Cashflow

Cashflow is the lifeblood of the business and more businesses fail through lack of cash than for any other reason. Cash is generated through the conversion of work into debtors and then into payment and also through the deferral of the payment of supplies for as long a period that can be negotiated.

The objective must be to keep stock, work in progress and debts to a minimum and creditors to a maximum. Trends in important ratios as well as absolute values can help in assessing the business performance.

Debtor days

This is calculated by dividing your trade debtors by annual sales and multiplying by 365. It shows the number of days' credit being afforded to your customers and should be compared both with your normal trade terms and the previous month's figures. Normal procedures should involve the preparation of a monthly-aged list of debtors showing the name of the customer, the value and to which month it relates.

The oldest and largest debtors can be seen at a glance for immediate consideration of what further recovery action is needed. The list may also show over-reliance on one or two large customers or the need to stop supplying a particularly bad payer until his arrears have been reduced to an acceptable level. Consideration should be given to making up bills to a date before the end of the month and making sure the accounts are sent out immediately, followed by a statement four weeks later.

Consider giving discounts for prompt payment. If all else fails, and legal action for recovery is being contemplated, call at the County Court and ask for their leaflets, numbers 1 to 4.

Stockturn

The level of stock should be kept to a minimum and the number of days' stock can be calculated by dividing the stock by the annual purchases and multiplying by 365. A worsening trend on a month-by-month basis shows the need for action. It is important to make regularly a full inventory of all stock and dispose of old or surplus items for cash. A stock control procedure to avoid stock losses and to keep stock to a minimum should be implemented.

RUNNING THE BUSINESS

Profitability

Whilst cash is vital in the short-term, profitability is vital in the medium-term. The two key percentage figures are the gross profit percentage and the net profit percentage. Gross profit is calculated by deducting the cost of materials and direct labour from the sales figures whilst net profit is arrived at after deducting all overheads.

Possible reasons for changes in the gross profit percentage are:

- not taking full account of increases in materials and wages in the pricing of jobs

- too generous discount terms being offered

- poor management, overmanning, waste and pilferage of materials

- too much down-time on plant which is in need of replacement.

If net profit is deteriorating after the deduction of an appropriate reward for your own efforts, including an amount for your own personal tax liability, you should review each item of overhead expenditure in detail asking the following questions:

- can savings be made in non-productive staff?

- is sub-contracting possible and would it be cheaper?

- have all possible energy-saving methods been fully explored?

- do the company's vehicles spend too much time in the yard; can they be shared and their number reduced?

- is the expenditure on advertising producing sales - review in association with 'marketing' above?

Over-trading

Many inexperienced businessmen imagine that profitability equals money in the bank. In some cases, particularly where the receipts are wholly in cash, this may be the case. Often increased business means higher stock inventories, extra wages and overheads, increased capital expenditure on premises and plant all of which require short-term finance.

OVER-TRADING

Additionally, if the debtors show a marked increase as the turnover rises, the proprietor may find to his surprise that each expansion of trade reduces rather than increases his cash resources and he is continually having to rely on extensions to his existing credit. The business, which had enough funds for start-up, finds it does not have sufficient cash to run at the higher level of operation and the bank manager may be getting anxious about the increasing overdraft.

It is essential for those who run a business which operates on credit terms to be aware that profitability does not necessarily mean increased cash availability. Regular monthly management information on marketing and finance as described in this chapter will enable over-trading to be recognised and remedial action to be taken early.

If the situation is appreciated only when the bank and other creditors are pressing for money, radical solutions may be necessary such as bringing in new finance, sale and leaseback of premises, a fundamental change in the terms of trade or even selling out to a buyer with more resources. Professional help from the firm's accountant will be needed, whilst the Business Link has counsellors experienced in advising on both the marketing and financial aspects of such situations.

Break-even point

The costs of a business may be divided into two types - variable and fixed.

Variable costs are those which increase or decrease as the volume of work goes up or down and include such items as materials used, direct labour and power machine tools.

Fixed costs are not related to turnover and are sometimes called fixed overheads. They include rent, rates, insurance, heat and light, office salaries and plant depreciation. These costs are still incurred even though few or no sales are being made.

Many small businessmen run their enterprises from home using family labour as back-up; they sell mainly their own labour and buy materials and hire plant only as required. By these means they reduce their fixed costs to a minimum and start making profits almost immediately. However, larger firms which have business premises, perhaps a small workshop, an office and vehicles, need to know how much they have to sell to cover their costs and become profitable.

In the case of a new business it is necessary to estimate this figure but where annual accounts are available a break-even chart based on them can be readily prepared.

RUNNING THE BUSINESS

Suppose the real or estimated figures (expressed in £000s) are:

	%	£
Sales	100	400
Variable costs	66	265
Gross profit	34	135
Fixed costs	13	50
Net Profit	21	85

Break-even point = 50 ÷ 1 - variable costs / sales

= 50 ÷ (1 - 0.6625)

= 50 ÷ 0.3375

= £148 (thousand)

In practice things are never quite as clear-cut as the figures show, but nevertheless this is a very useful tool for assessing not only the break-even point but also the approximate amount of loss or profit arising at differing levels of turnover and also for considering pricing policy.

Chapter 3
Taxation

INCORPORATION

The first decision usually required to be made from a taxation point of view is which trading entity to adopt. The options available are set out below.

Sole traders

A sole trader is a person who is in business on his own account. There is no statutory requirement to produce accounts nor is there a necessity to have them audited. A sole trader may, however, be required to register for PAYE and VAT purposes and maintain records so that income tax and VAT Returns can be made. A sole trader would be personally liable for all the liabilities of his business.

Partnership

A partnership is a collection of individuals in business on their own account and whose constitution is generally governed by the Partnership Act 1890. It is strongly recommended that a partnership agreement is also established to determine the commercial relationship between the individuals concerned.

The requirements in relation to accounting records and returns are similar to those of sole traders and in general a partner's liability is unlimited.

Limited Company

This is the most common business entity. Companies are incorporated under the Companies Act 1985 which requires that an annual audit is carried out for all companies with a turnover in excess of £350,000 and that accounts are filed with the Companies Registrar. Companies with a turnover of between £90,000 and £350,000 are required to have their accounts examined and an accountant's report attached. If the turnover does not exceed £90,000 no audit is required. Generally an individual shareholder's liability is limited to the amount of the share capital he is required to subscribe.

TAXATION

Advantages

In view of the problems and costs of incorporating an existing business, it is important to try and select the correct trading medium at the commencement of operations. It is not true to say that every business should start life as a company. Many businesses are carried on in a safe and efficient manner by sole traders or partnerships. Whilst recognising the possible commercial advantages of a limited company, taxation advantages exist for sole traderships and partnerships such as income tax deferral and National Insurance saving. No decision should be taken without first seeking professional advice.

The incorporation of a business or the immediate formation of a limited company is less attractive that it used because the maximum rate of taxation is only 40%, especially when considering the high levels of Class 1 National Insurance compared to those contributions required under Class 2 and Class 4. Taking these factors into account, it is suggested that taxable profits per partner need to be in excess of £48,340 per annum in order for it to advantageous from a tax point of view for a limited company to be formed.

The benefit of limited liability should not be ignored although this can largely be negated by banks seeking personal guarantees. In addition, it may be easier for the companies to raise finance because the bank can take security on the debts of the company which could be sold in the future, particularly if third-party finance has been obtained in the form of equity.

Self-employed sole traders and partnerships

For those persons deciding to trade on their own account or in partnership a new basis of charging tax on business profits took effect from 6 April 1994 for new businesses and from 6 April 1996 for businesses already established at 6 April 1994.

For new businesses there may still be an advantage in having an accounting date early in the next tax year, that is shortly after 6 April. The advantage, however, is less that it would have been under the old rules and professional advice will be needed to choose the best accounting date for tax purposes.

For businesses existing at 6 April 1994 the way income tax is charged on business profits is to change radically. For the 1997/98 tax year onwards, profits assessed for tax in a given tax year will be the profits of the accounting period ending in that tax year and not as now those of the preceding year. For example, accounts ended 30 April 1997 will be assessed 1997/98 (the first year of the current year basis). Profits for the year ended 30 April 1994 will be assessed 1995/96 (the last year on the prior-year basis).

PARTNERSHIPS

Transitional rules will apply for the tax year 1996/97 whereby the profits of two accounting periods will be used to give an average. For example, for a self-employed business with a 30 April year end, the average period began on 1 May 1994 and will end on 30 April 1996.

Due to the averaging of profits charged to tax, a tax planning opportunity will arise subject to anti-avoidance provisions introduced by the Inland Revenue to penalise the artificial movement of profit into the averaging period.

In particular, the following areas may offer tax planning opportunities:

- the purchase as opposed to the leasing of capital equipment to be used in the business

- the introduction of a partner into the business

- where a partnership is considering, for sound business reasons, re-financing by way of personal loans rather than a partnership loan.

Professional advice will be needed to take the maximum advantage of these opportunities.

Self-assessment

Together with changes in the way taxable profits will be measured, the Inland Revenue are introducing a new system of charging and collecting personal tax. Again from the tax year 1996/97 the burden of assessing tax will shift from the Inland Revenue to the individual tax payer. The main features of this new system are as follows:

- the onus is on the taxpayer to provide information and complete returns

- tax will be payable on different dates

- the taxpayer will have a choice: he can calculate and pay his tax liability at the same time as making his return and this will need to be done by 31st January following the end of the tax year. Alternatively, he can send in his tax return much earlier and the Inland Revenue will calculate the tax to be paid on 31 January

- the important aspect to the new system is that if the return is late, or the tax is paid late, there will be automatic penalties imposed on the taxpayer.

TAXATION

Assessments

To businessmen, income tax assessments are an anathema. They should resist the temptation to tear them up or put them behind the clock and forget about them. All assessments should be checked for accuracy immediately and, if excessive, the instructions on the notice about making an appeal should be followed and the appeal sent to the Tax District that issued the assessment. The appeal should also show how much of the tax charged should be postponed because the assessment is too high.

If this in not done within 30 days of the issue of the notice, the assessment becomes final and the Inspector (and the General Commissioners if they are asked to adjudicate) may well not accept a late appeal. In this event the taxpayer has no alternative but to pay the tax as charged on the assessment which may be estimated or contain additions to the profits which have not been agreed by him and his accountant. If the appeal is to be made by the accountant check with him before the 30 days are up to ensure that it has been submitted.

Keep copies of all correspondence with the Inspector and Collector. Letters can be mislaid or fail to be delivered and it is essential to have both proof of what was sent as well as a permanent record of all correspondence.

Dates tax due

Income Tax 1996/97

Earned income (such as trading profits) 50% on 1 January 1997 and a further 50% on 1 July 1997. Unearned income (such as rents and interest) due on 1 January 1997.

Capital Gains Tax

Tax on gains in year ended 5 April 1996 is due on 1 December 1996 (or within 30 days of issue of the notice of assessment). Tax on gains arising in the current year are due on 1 December 1997.

Tax in business

Spouses in business

If spouses work in the business, perhaps answering the phone, making appointments, writing business letters, making up bills and keeping the books, they should be properly remunerated for it. Being a payment to a family member the Inspector of Taxes will be understandably cautious in allowing remuneration in full as a business expense. The payment should be:

TAX IN BUSINESS

- actually paid to them, preferably weekly or monthly and in addition to any housekeeping monies

- recorded in the business book

- reasonable in amount in line with their duties and the time spent on them.

If the wages paid to them exceed £60.99 p.w., Class 1 employer's and employee's NIC becomes due and if they exceed £3,765 p.a. (assuming they have no other income) PAYE tax will also be payable.

It should also be noted that once small businesses are well established and the spouse's earnings are approaching the above limits, consideration may be given to bringing them is as a partner. This has a number of effects:

- there is a lesser need to relate the spouse's income (which is now a share of the profits) to the work they do

- they will pay Class 2 and Class 4 NIC instead of the more costly Class 1 contributions and PAYE will no longer apply to their earnings

- but remember that as partners, they have unlimited liability.

Premises

Many small businessmen cannot afford to rent or buy commercial premises and run their enterprises from home using part of it as an office where the books and vouchers, clients' records and trade manuals are kept and estimates and plans are drawn up. In these circumstances, a portion of the outgoings on the property may be claimed as a business expenses. Professional advice should be sought such that the capital gains tax exemption which applies on the sale of the main residence is not lost.

Vehicles

Car expenses for sole traders and partners are usually split on a fractional mileage basis between business journeys, which are allowable, and private ones, which are not, and a record of each should be kept. If the business does work only on one or two sites for only one main contractor the inspector may argue that the true base of operations is the work site not the residence and seek to disallow the cost of travel between home and work. It is tax-wise and sound business practice to have as many customers as possible and not work for just one client.

TAXATION

Business entertainment

No tax relief is due for expenditure on business entertainment and neither is the VAT recoverable on gifts to customers, whether they are from this country or overseas. However, the cost of small trade gifts not exceeding £15 per person in value is still admissible provided that the gift advertises the business and does not consist of food or tobacco.

Income tax

Personal Allowances

The current personal allowance for a single person is £3,765. The personal allowance for people aged 65 to 74 and over 75 years are £4,910 and £5,090 respectively. The married couple's allowance is £1,790 and £3,115 for a couple between the ages of 65 and 74 and £3,155 for a couple over 75 years.

The value of the married person's allowance for 1996/97 is the same as 1995/96 at 15%.

Taxation of husband and wife

The married persons' allowance is initially due to the husband, but if his income is too small to use it all he may transfer the surplus to his wife. Once he has passed some allowances in this way he cannot change his mind and ask for them back. Married couples may choose how they wish the allowance to be allocated between them, subject to the right of the wife to claim half the allowance if she wishes.

A married women is treated in much the same way as a single person with her own personal allowance and basic rate band. Husband and wife each make a separate return of their own income and the Inland Revenue deals with each one in complete privacy; letters about the husband's affairs will be addressed only to him and about the wife's only to her (unless the parties indicate differently).

Rates of tax

The rates of tax for 1996/97 are as follows:

> Lower rate: 20% on taxable income up to £3,900
> Basic rate: 24% on taxable income between £3,900 and £25,500
> Higher rate: 40% on taxable income over £25,500

RATES OF TAX

Mortgage interest relief

The ceiling for relief is unchanged for 1996/97 at £30,000 as is the value of relief at 15%. If the loan is in the name of one spouse only, that one gets all the relief due. If it is in the names of husband and wife, it is allowed equally between them However, if both agree, they can *jointly* ask for relief to be divided in any proportion that they wish.

Business losses

These are allowed only against the income of the person who incurs the loss. For example, a loss in the husband's business cannot be set against the wife's income from employment.

Joint income

In the case of joint ownership by husband and wife of assets which yield income, such as bank and building society accounts, shares and rented property, the Revenue will treat the income as arising equally to both and each will pay tax on one half of the income. If, however, the asset is owned in unequal shares or one spouse only and the taxpayer can prove this, then the shares of income to be taxed can be adjusted accordingly if a joint declaration is made to the tax office setting out the facts.

Generally

Special rules apply in the year of marriage or separation and divorce and on the death of the husband or wife. Contact your accountant or local tax office for information.

Capital Gains Tax

Where an asset is disposed of, the first £6,300 of the gain is exempt from tax. In the case of husbands and wives, each has a £6,500 exemption so if the ownership of the assets is divided between them, it is possible to claim exemption on gains up to £12,600 jointly in the tax year. Any remaining gain is chargeable as though it were the top slice of the individual's income; therefore according to his or her circumstances it might be charged at 20%, 24% or 40%.

TAXATION

Retirement relief may be due on the disposal of business assets after the age of 50 for disposals on or after 28 November 1995 (or before that date where retirement is due to proven ill health) - for disposals before 28 November 1995, the relevant age was 55. The maximum relief against capital gains is £250,000 plus one half of the gains between £250,000 and £1,000,000. A businessman contemplating retirement, or sale of business when aged 50 or over, should consult his accountant *before* taking any steps and *before* changing his working pattern (e.g. going part time).

Self-employed NIC rates (from 6 April 1994)

Class 2 rate

Charged at £6.05 per week. If earnings are below £3,430 p.a averaged over the year, ask the DSS about 'small income exception'. Details are in leaflet NI 27A.

Class 4 Rate

Business profits up to £6,860 p.a are charged at NIL. Profits between £6,860 and £23,660 p.a. are charged at 6% of the profit. There is no charge on profits over £23,660 p.a. so the maximum amount of Class 4 contributions is £1,008.00. Class 4 contributions are collected by the Inland Revenue along with the income tax due.

Corporation Tax (years ended 31 March 1995 and 31 March 1996)

For the year ending 31 March 1996, Corporation Tax is charged at 25% for profits up to £300,000. This rate is reduced to 24% on profits up to £300,000 for the year ending 31 March 1997. Where the accounting date of a company is not 31 March, profits have to be apportioned on a time basis to the respective tax years. Profits exceeding £300,000 will be effectively charged at 35% for 1996 and 35.25% for 1997 up to £1,500,000 when the rate reduces to 33%. Companies can carry back trading losses for up to 3 years.

Capital allowances (depreciation) rates

Plant and machinery:	25%
Business motor cars - cost up to £12,000:	25%
- cost over £12,000:	£3,000 (maximum)
Industrial buildings:	4%
Commercial and industrial buildings in Enterprise Zones:	100%

THE CONSTRUCTION INDUSTRY TAX DEDUCTION SCHEME

THE CONSTRUCTION INDUSTRY TAX DEDUCTION SCHEME

General

The Construction Industry Tax Deduction Scheme is knowen universally as the '714' scheme after the number of the official form around which the whole system revolves. The Government is proposing important changes to the scheme from the 31 July 1998 which will require sub-contractors to achieve a minimum turnover before a Certificate will be granted. As part compensation, contractors will be able to deduct tax at a lower rate so that the tax deducted, equals the annual liability.

As the scheme operates whenever a contractor makes a payment to a sub-contractor, the businessman should visit his local income tax enquiry office and obtain copies of the Revenue booklet IR 14/15 and leaflet IR 40 which explain the conditions under which the Revenue will issue a 714 certificate and precisely when the scheme applies.

The 714 Certificate

There are four types:

    714I - which is issued to individuals

    714P - which is issued to partners

    714C - which is issued to most limited companies

    714S - which is issued as explained below.

If the sub-contractor does not hold a valid tax certificate (714I, 714P, 714C or 724S) issued to him by the Inland Revenue, then the contractor *must* deduct 25% tax from the whole of any payment made to him (excluding the cost of any materials) and to account to the Revenue for all amounts so withheld. To enable the sub-contractor to prove to the Inspector of Taxes that he has suffered this tax deduction the contractor must give him a certificate on form SC60 showing the amount withheld.
    These SC60 forms must be carefully filed for production to the Inspector after the end of his accounting year along with the business profit and loss account and balance sheet. Any tax deducted in this way over and above the sub-contractor's agreed liability for the year will be repaid by the Inland Revenue.

TAXATION

If, however, he holds a 714 certificate the payment may be made in full without deducting tax (in the case of a 714S there is a weekly limit after which tax is deductible).

A business is not obliged by law to seek an exemption certificate and can legally work in the construction industry without one. As a matter of practice, however, many main contractors are reluctant to undertake the additional paperwork required when tax has to be deducted and accounted for to the Revenue and will give work to a sub-contractor with a 714 certificate in preference to one without.

A small business that does work only for the general public and small commercial concerns, is outside the scheme and does not need a 714 certificate to trade. If, however, it engages other contractors to do jobs for it, the business would have to register under the scheme as a contractor and deduct tax from any payment made to a sub-contractor who did not produce a valid 714 certificate.

VAT

The general rule about liability to register for VAT is given in the VAT office notes. It is possible to give here only a brief outline of how the tax works. The rules which apply to the construction industry are extremely complex and all traders must study *The VAT Guide* and other publications.

Registration for VAT is required if:

- at the end of any month the value of taxable supplies in the last 12 months exceeds the annual threshold

or

- there are reasonable grounds for believing the value of the taxable supplies in the next 30 days will exceed the annual threshold.

Taxable supplies include any zero-rated items. The annual threshold from 29 November 1995 is £47,000. The amount of tax to be paid is the difference between the VAT charged out to customers *(output tax)* and that suffered on payments made to suppliers for goods and services *(input tax)* incurred in making taxable supplies. Unlike income tax there is no distinction in VAT for capital items so that the tax charged on the purchase of, for example, machinery, trucks and office furniture, will normally be reclaimable as Input Tax.

VAT is payable in respect of three monthly periods known as 'tax periods'. You can apply to have the group of tax periods which fits in best with your financial year. The tax must be paid within one month of the end of each tax period. Traders who receive regular repayments of VAT can apply to have them monthly rather than quarterly. Not all types of goods and services are taxed at 17.5% (ie at the standard rate). Some are exempts and others are zero-rated.

At Butterworth-Heinemann we are determined to provide you with a quality service. To help supply you with information on relevant titles as soon as it is available, please fill in the form below and return to us using the FREEPOST facility. Thank you for your help and we look forward to hearing from you.

What title have you purchased? _____
Where was the purchase made ? _____
When was the purchase made? _____
Name (Please Print): _____
Job Title: _____
Street: _____
Town: _____
County: _____ Postcode: _____
Country: _____ Telephone: _____
Company Activity: _____
Signature: _____ Date: _____

* Please arrange for me to be kept informed of other books, journals and information services on this and related subjects (* delete if not required). This information is being collected on behalf of Reed International Books Ltd and may be used to supply information about products produced by companies within the Reed International Books group.

(FOR OFFICE USE ONLY)

BUTTERWORTH
HEINEMANN

Butterworth-Heinemann Limited – Registered Office: Michelin House, 81 Fulham Road, London, SW3 6RB. Registered in England 194771. VAT number GB: 340 242992

(For cards outside the UK please affix a postage stamp)

Direct Mail Department
Butterworth-Heinemann
FREEPOST
OXFORD
OX2 8BR
UK

VAT

Zero-rated

This means that no VAT is chargeable on the goods or services, but a registered trader can reclaim any *input* tax suffered on his purchases. For instance a builder pays VAT on the materials he buys to provide supplies of constructing but if he is constructing a new dwelling house, this is zero rated. He may reclaim this VAT or set it off against any VAT due on standard rated work.

Exempt

Supplies which are exempt are less favourably treated than those which are zero-rated. Again no VAT is chargeable on the goods or services but the trader cannot reclaim any *inpunput* tax suffered on his purchases.

Standard-rated

All work which is not specifically stated to be zero rated or exempt is standard-rated i.e VAT is chargeable at the current rate of 17.5% and the trader may deduct any *input* tax suffered when he is making his return to the Customs and Excise.

If for any reason a trader makes a supply and fails to charge VAT when he should have done so (eg mistakenly assuming the supply to be zero rated), he will have to account for the VAT himself out of the proceeds. If there is any doubt about the VAT position, it is safer to assume the supply is standard rated, charge the appropriate amount of VAT on the invoice and argue about it later.

Time of supply

The *time* at which a supply of goods or services is treated as taking place is important and is called the 'tax point' VAT must be accounted for to the Customs & Excise at the end of the accounting period in which this 'tax point' occurs. For the supply of *goods* which are 'built on site' the 'basic tax point' is the date the goods are made available for the customer's use, whilst for *services* it is normally the date when all work except invoicing is completed.

However, if you issue a tax invoice or receive a payment *before* this 'basic tax point' then that date becomes a tax point. In the case of contracts providing for stage and retention payments the tax point is either the date the tax invoice is issued or when payment is received, whichever is the earlier.

All the requirements apply to sub-contractors and main contractors and it should be noted that, when a contractor deducts income tax from a payment to a sub-contractor (because he has no valid 714), VAT is payable on the full gross amount *before* taking off the income tax .

TAXATION

Annual accounting

It is possible to account for VAT other than on a specified three month period. Annual accounting provides for nine equal instalments to be paid by direct debit with annual return provided with the tenth payment.

Note: Customs & Excise have announced changes to the VAT Annual Accounting Scheme which will be introduced on 1 April 1996. The new scheme will continue to be voluntary and will particularly benefit businesses with turnover of below £100,000, although the scheme remains available to businesses with an annual turnover of less than £300,000.

Cash accounting

If turnover is below a specified limit, currently £350,000, a tax payer may account for VAT on the basis of cash paid and received. The main advantages are automatic bad debt relief and a deferral of VAT payment where extended credit is given.

Bad debts

Relief is available for debts over 6 months.

Chapter 4
Estimating

Pity the poor estimator! If a job goes well there will be a queue of agents, foremen, tradesmen, surveyors and buyers to take the credit. But if a job loses money, the one person who will be left isolated to take the blame is the unsung hero of the construction industry - the estimator!

His art is highly dependent upon making a series of intelligent guesses to fill in the gaps of information not covered by the specification, drawings and/or the bills of quantities. The quality of these guesses (or subjective judgements as the jargon would have it!) will often make the difference between a job making a profit or a loss.

It is possible, of course, to have too much information! There are many examples of a contractor carrying out the first phase of a contract but losing the second phase in open tender despite the fact of having the site set-up already there. The reason for the loss is usually due to having too much local knowledge gained from working on Phase 1 and including the cost of the known local risks in the second tender. Ignorance frequently prevails!

Apart from determining the contract sum at the tender stage, a properly prepared estimate can also be used for the following purposes:

1. calculation of bonus targets

2. preparation of material schedules

3. analysis of anticipated and actual costs

4. production of a programme

5. preparation of monthly valuations and the final account.

This book, however, is concerned only with the preparation of estimates at the tender stage. It should be recognized that estimating is a very imprecise art. This often surprises people outside the construction industry who imagine it to be '... merely a matter of counting bricks and pricing them' as someone once said to me. If only it was!

ESTIMATING

Even taking this comment at face value shows how inaccurate this view is. There are many different kinds of bricks all with a different value. The cost of sand and cement for the mortar varies from supplier to supplier as does the hire cost of a mixer. The labour costs are the most likely to show the greatest variation.

No two men work at the same pace and produce the same volume of work. So even an uncomplicated task such as building a wall can produce significantly different estimates. When it comes to more complicated work, the likelihood of estimates showing wide variations in value increases proportionately.

The main divisions in a properly prepared estimate are:

1. Own work which can be sub-divided into

    (a) labour
    (b) materials
    (c) plant

2. work to be sub-let to sub-contractors

3. site overheads

4. office overheads

5. profit.

When an enquiry is received, a contractor should decide very quickly which parts of the work he would sub-let if he was awarded the contract. He should then send the relevant extracts from the enquiry documents (specification/bills of quantities/drawings) to two or three sub-contractors whilst he is pricing the remainder of the estimate.

Labour

An experienced contractor should know the net charge-out rate of the men he directly employs. The basic labour rate on which the information in this book is based is set out at the beginning of Chapter 5. Every contractor will probably have his own 'customized' version but the main items such as NHI contributions, overtime payments, bonuses, etc., must be included.

The particular circumstances of the job must also be considered. The work may have to be carried out outside normal trading hours or done in unpleasant working conditions - both these situations will produce higher labour costs.

MATERIALS

Materials

Although it is usually possible to obtain a better price from another materials supplier, a contractor should weigh up whether it is worth spending a lot of his time investigating other sources if the savings are marginal.

A supplier who delivers on time (including the occasional small item when necessary) and replaces defective materials without query is probably worth supporting even if his prices are slightly higher than those of some of his competitors. It is still worth checking other prices now and then, but reliability and service have a real value and should be recognized.

The most difficult aspect of pricing materials is making a realistic assessment for waste and theft. Materials which are bought in bulk, but are only part of the order allocated to a particular job, usually attract the highest waste percentage which can be as high as 15 to 20%. Specialized 'one-off' items such as cylinders, heaters and the like are usually better looked after and the waste factor could be as low as 1%.

Theft is another problem and a prudent contractor should make some allowance in his estimates to cover for this frustrating part of the industry.

Plant

Most small contractors hire in plant as necessary so the estimator need only assess the time the plant will be required because the hire rate will be easily established. One point to watch out for is the question of minimum hire periods.

Even if a piece of equipment is only required for one hour, the full cost of a day's hire must be included in the rates if that is the minimum hire period that can be obtained.

Site overheads

The range of overheads to be provided on site will vary widely depending on the size and nature of the job. There are two main types of overheads - fixed and time-related. Examples of these will clearly reveal the difference between them.

Fixed overheads are those of a non-recurring nature such as the establishment of the site huts. The cost of this operation - say £500 - must be allowed for whether the job lasts 4 weeks or 104 weeks and is a fixed sum.

Time-related overheads, however, are totally linked to the time the facility or service is required. In the case of site hutting the estimator should allow a weekly cost for the anticipated contract period plus any time that will be needed for carrying out maintenance work.

The following should act as a checklist for both fixed and time-related overheads:

ESTIMATING

site supervision
site accommodation
lighting
water
safety, health and welfare
removal of rubbish
cleaning
drying out
protection of work
security
small tools
insurances
travelling time and fares
scaffolding
temporary services
temporary fencing and screens

Office overheads

All contractors should give careful thought to the cost of their office overheads and in particular should be aware of the relationship, expressed as a percentage, between these costs and turnover.

The following is an example of a small contractor's overheads. This list is not exhaustive.

			£
Rent		(52 x £50)	2,600
Printing, stationery, etc.			150
Telephone		(12 x £50)	600
Van	HP	(12 x £200)	2,400
	Insurance		400
	Tax		100
	Petrol	(52 x £30)	1,560
	Repairs say		500
Carried forward			8,310

OVERHEADS

		£
Brought forward		8,310
Part-time clerk/typist (52 x 30 hours x £3.50)		5,460
Photocopier - lease	(52 x £25)	1,300
Word-processor - lease	(52 x £20)	1,040
Advertising	(52 x £20)	1,040
Membership of professional body		150
Sundries		1,000
		18,300

Let us assume that the firm's turnover (that is, the total amount billed for the previous year) is £100,000. A simple calculation of £18,300 x 100 ÷ £100,000 shows that the relationship of overheads to turnover is 18.3%. I have set out a table below showing how this percentage changes when the level of the overheads or the turnover changes.

Overheads £	Turnover £	%
18,300	100,000	18.30
18,300	80,000	22.90
18,300	120,000	15.25
24,000	100,000	24.00
30,000	80,000	37.50
12,000	120,000	10.00

It is important to know the current percentage so it can be added to each quotation (although not necessarily shown). Ask your accountant for the cost of the overheads for last year to assess the percentage to be added to the costs. Unless something dramatic happens mid-year this figure needs only be examined annually. An example which would affect the overheads total could be the

ESTIMATING

employment of a non-working supervisor or the purchase of an extra vehicle. A cancellation of a project or an unexpected increase in work would also affect the turnover figure.

It can be seen, therefore, that although working on a notional percentage is acceptable you should keep your eye open for any circumstances which may make a significant alteration to it.

Profit

Profit is the difference between income and cost. Some newly established contractors find it puzzling that although they are certain that their work is profitable, it is not reflected in the bank balance! The simple explanation for this apparent paradox is that as soon as profit is earned it is used to finance the next stage of work or the next job. It only appears as a tangible asset when either trading stops or after a series of profitable jobs are completed and paid for.

Expansion of the business (and it is very difficult for a successful contractor not to expand) will further delay the surfacing of the profit and even a well-planned expansion programme will usually result in an increased overdraft. The profit is still in the business but not in the form of cash in the bank but in debtors' accounts and work in progress.

Chapter 5
Rates for Measured Work

Generally

The rates contained in this chapter apply to contracts ranging from minor works to those up to £50,000 in value. The rates are exclusive of VAT which may be chargeable depending on the nature of the work and the status of the contractor. These rates are based upon the national average and the following regional adjustments should be made:

Scotland	+	4%
Wales		-
Northern Ireland	-	18%
England		
South West	-	1%
South East	-	7%
Home Counties	-	7%
Inner London	+	5%
Outer London	-	2%
East Anglia	-	8%
Midlands	+	2%
North West		-
North East	+	2%

Item descriptions

These contain brief descriptions of items. Each item includes all the work normally associated with that particular item even if it is not expressly stated.

Unit

This column gives the unit in which the item is normally measured.

RATES FOR MEASURED WORK

Labour hours

The time taken for the fixing of each of the items is expressed as a decimal fraction of an hour; therefore 0.50 hours equals thirty minutes.

Net labour

This column gives the total labour cost of the item concerned. Although the 'Labourer' (or Adult General Operative) rate is calculated overleaf, the amount stated is based upon the 'Roofer' rate only. The rates are based upon the rates agreed and published by the National Joint Council for the Building Industry and are effective from 26 June 1995.

Craftsman	£
Hours worked 47.8 weeks at £173.55	8,295.69
Sick pay, say 1 week at £173.55	173.55
Non-productive overtime say 1 week at £173.55	173.55
Bonus, say 48 weeks at £25.00 per week	1,200.00
National Health Insurance 7% x £8,295.69	580.70
Holiday pay, 48 stamps at £19.30	_926.40_
Severance pay 1% x £11,349.89	113.50
Redundancy pay 1% x £11,349.89	113.50
Employer's liability and third-party insurance 2% x £11,349.89	227.00
CITB levy 0.25% x £11,349.89	_28.38_
£	_11,832.27_
Cost per hour: £11,832.27 ÷ 1,801.80 hours £	_6.57_

RATES FOR MEASURED WORK

Labourer £

Hours worked 47.8 weeks at £143.52	6,860.26
Sick pay, say 1 week at £143.52	143.52
Non-productive overtime, say 1 week at £143.52	143.52
Bonus, say 48 weeks at £15.00 per week	720.00
National Health Insurance 7% x £6,860.26	480.22
Holiday pay, 48 stamps at £19.30	926.40
£	9,273.92
Severance pay 1% x £9,273.92	92.74
Redundancy pay 1% x £9,273.92	92.74
Employer's liability and third-party insurance 2% x £9,273.92	185.48
CITB levy 0.25% x £9,273.92	23.18
£	9,668.06

Cost per hour: £9,668.06 ÷ 1,801.80 hours £ 5.37

There are many different types of roofing covered by this book and the hourly labour rate has been taken as £7.00.

Net material

Generally speaking, the costs reflect manufacturers' list prices and exclude trade discounts which are usually negotiated by individual contractors depending upon creditworthiness, turnover and length of association with the supplier. The prices are based upon the purchase of manufacturers' standard packs.

RATES FOR MEASURED WORK

Overheads/profit

Overhead charges are those costs which are not directly linked to any particular contract. They are usually expressed as a percentage of the net cost of labour and materials and are calculated by dividing the total projected annual overhead cost by the projected net turnover. They are explained in more detail in Chapter 4. The percentage addition for overheads and profit is taken at 15%.

Total

The total column is the sum of the net labour, net material and overheads/profit columns.

Builder's work

It should be noted that no allowances have been made for scaffolding but information on the cost of hiring it can be found in Chapter 6.

Sundries

Where the work involves stripping roofs and removing debris the following costs should be applied for loading into skips.

Assuming that skips would be filled to capacity the costs should be:

Size of skip m3	Daily hire rate £	Cost per m3 £
3	28.00	9.33
5	38.00	7.60
6	50.00	8.33
8	54.00	6.75

The cost of loading into barrows and wheeling, say 25m, is taken separately at 1.2 man-hours at £7.00 per m3 = £8.40 per m3. This figure should have 15% added for profit and overheads to produce an overall charge-out rate of £9.66.

RATES FOR MEASURED WORK

The cost of final disposal is:

Size of skip m3	Cost per m3	Profit/ overheads 15%	Total £	Loading £	Total £
3	9.33	1.40	10.73	9.66	20.39
5	7.60	1.14	8.74	9.66	18.40
6	8.33	1.25	9.58	9.66	19.24
8	6.75	1.01	7.76	9.66	17.42

Loading into lorries

It is assumed that the rubbish can be barrowed into the lorries via ramps and walkboards. It is also assumed that the lorry can remove three full loads of 8m3 per day and that the hire cost of a lorry and driver is £28.00 per hour.

Lorry and driver 8 hours at £28.00   £224.00
Tipping charge: 3 at £10.00   30.00

                               £254.00

divided by 24m3   =   £10.58 per m3

Removal cost m3 £	Profit/ overheads £	Total £
10.58	1.59	12.17

Generally

There are many variable factors affecting the decision whether to have a skip or a lorry and each must be carefully judged individually depending upon the weather, the quantity to be moved, the labour available to load and other relevant items.

MATERIAL COSTS

	Unit	Material supply (£)	Material waste (%)	Total (£)

H31 METAL PROFILED SHEETING

Aluminium alloy roll-formed profiled sheets (Precision Metal Forming Ltd)

Cladding, 0.70mm thick, plain or stucco

profile 13.5/3	m2	4.70	5.0	4.94
profile 19	m2	4.37	5.0	4.59
profile MS20	m2	4.66	5.0	4.89
profile 32	m2	4.85	5.0	5.09
profile 35	m2	5.18	5.0	5.44
profile 38A	m2	5.09	5.0	5.34
profile 40	m2	4.66	5.0	4.89
profile 46	m2	5.18	5.0	5.44
profile 60	m2	5.82	5.0	6.11

Cladding, 0.90mm thick, plain or stucco

profile 13.5/3	m2	5.93	5.0	6.23
profile 19	m2	5.51	5.0	5.79
profile MS20	m2	5.88	5.0	6.17
profile 32	m2	6.12	5.0	6.43
profile 35	m2	6.53	5.0	6.86
profile 38A	m2	6.43	5.0	6.75
profile 40	m2	5.88	5.0	6.17
profile 46	m2	6.53	5.0	6.86
profile 60	m2	7.35	5.0	7.72
profile 100	m2	8.40	5.0	8.82

Cladding, 1.20mm thick, plain

profile 13.5/3	m2	7.71	5.0	8.10
profile 19	m2	7.16	5.0	7.52
profile MS20	m2	7.63	5.0	8.01
profile 32	m2	7.96	5.0	8.36
profile 35	m2	8.48	5.0	8.90
profile 38A	m2	8.36	5.0	8.78

MATERIAL COSTS

Aluminium sheeting (cont'd)	Unit	Material supply (£)	Material waste (%)	Total (£)
profile 40	m2	7.63	5.0	8.01
profile 46	m2	8.48	5.0	8.90
profile 60	m2	9.54	5.0	10.02
profile 100	m2	10.91	5.0	11.46

Cladding, 0.70mm thick, AD80/ standard backing coat

profile 13.5/3	m2	8.36	5.0	8.78
profile 19	m2	7.76	5.0	8.15
profile MS20	m2	8.28	5.0	8.69
profile 32	m2	8.63	5.0	9.06
profile 35	m2	9.20	5.0	9.66
profile 38A	m2	9.06	5.0	9.51
profile 40	m2	8.28	5.0	8.69
profile 46	m2	9.20	5.0	9.66
profile 60	m2	10.46	5.0	10.98
profile 100	m2	11.83	5.0	12.42

Cladding, 0.90mm thick, AD80/ standard lacquer

profile 13.5/3	m2	10.37	5.0	10.89
profile 19	m2	9.64	5.0	10.12
profile MS20	m2	10.28	5.0	10.79
profile 32	m2	10.71	5.0	11.25
profile 35	m2	11.42	5.0	11.99
profile 38A	m2	11.25	5.0	11.81
profile 40	m2	10.28	5.0	10.79
profile 46	m2	11.42	5.0	11.99
profile 60	m2	12.85	5.0	13.49
profile 100	m2	13.64	5.0	14.32

Aluminium alloy pressed sheets

Cladding, 0.70mm thick, plain or stucco

profile PR8	m2	5.29	5.0	5.55
profile PM13	m2	5.66	5.0	5.94
profile PL19	m2	5.85	5.0	6.14
profile PG22	m2	6.21	5.0	6.52
profile PS47	m2	7.68	5.0	8.06

MATERIAL COSTS

	Unit	Material supply (£)	Material waste (%)	Total (£)
Cladding, 0.90mm thick, plain or stucco				
profile PR8	m2	6.68	5.0	7.01
profile PM13	m2	7.14	5.0	7.50
profile PL19	m2	7.38	5.0	7.75
profile PG22	m2	7.83	5.0	8.22
profile PS47	m2	9.69	5.0	10.17
Cladding, 1.20mm thick, plain				
profile PR8	m2	8.68	5.0	9.11
profile PL19	m2	9.60	5.0	10.08
profile PG22	m2	10.17	5.0	10.68
profile PS47	m2	12.59	5.0	13.22
Cladding, 0.70mm thick, AD80/ standard backing coat				
profile PR8	m2	9.42	5.0	9.89
profile PM13	m2	10.07	5.0	10.57
profile PL19	m2	10.41	5.0	10.93
profile PG22	m2	11.04	5.0	11.59
profile PS47	m2	13.66	5.0	14.34
Cladding, 0.90mm thick, AD80/ standard lacquer				
profile PR8	m2	11.69	5.0	12.27
profile PM13	m2	12.50	5.0	13.13
profile PL19	m2	12.92	5.0	13.57
profile PG22	m2	13.70	5.0	14.39
profile PS47	m2	16.95	5.0	17.80

Galvanized steel roll-formed profiled sheets

Cladding, 0.70mm thick

profile 13.5/3	m2	4.63	5.0	4.86
profile 19	m2	4.29	5.0	4.50
profile MS20	m2	4.59	5.0	4.82
profile 32	m2	4.78	5.0	5.02
profile 35	m2	5.09	5.0	5.34

MATERIAL COSTS

Steel sheeting (cont'd)	Unit	Material supply (£)	Material waste (%)	Total (£)
profile 38A	m2	5.02	5.0	5.27
profile 40	m2	4.59	5.0	4.82
profile 46	m2	5.09	5.0	5.34
profile 60	m2	5.73	5.0	6.02
profile 100	m2	6.55	5.0	6.88

Cladding, 0.90mm thick

profile 13.5/3	m2	5.83	5.0	6.12
profile 19	m2	5.41	5.0	5.68
profile MS20	m2	5.78	5.0	6.07
profile 32	m2	6.02	5.0	6.32
profile 35	m2	6.42	5.0	6.74
profile 38A	m2	6.31	5.0	6.63
profile 40	m2	5.78	5.0	6.07
profile 46	m2	6.42	5.0	6.74
profile 60	m2	7.21	5.0	7.57
profile 100	m2	8.24	5.0	8.65

Cladding, 1.20mm thick

profile 13.5/3	m2	7.55	5.0	7.93
profile 19	m2	7.01	5.0	7.36
profile MS20	m2	7.49	5.0	7.86
profile 32	m2	7.79	5.0	8.18
profile 35	m2	8.32	5.0	8.74
profile 38A	m2	8.18	5.0	8.59
profile 40	m2	7.49	5.0	7.86
profile 46	m2	8.32	5.0	8.74
profile 60	m2	9.36	5.0	9.83
profile 100	m2	10.69	5.0	11.22

Composite floor decking, 0.90mm thick

profile CF46	m2	6.75	5.0	7.09
profile CF51	m2	8.41	5.0	8.83
profile CF70	m2	6.27	5.0	6.58

MATERIAL COSTS

	Unit	Material supply (£)	Material waste (%)	Total (£)

Composite floor decking, 1.20mm thick

profile CF46	m2	8.76	5.0	9.20
profile CF51	m2	11.25	5.0	11.81
profile CF70	m2	8.11	5.0	8.52

Colour coated galvanized steel roll-formed profiled sheets

Cladding, 0.70mm thick, external face HP200, internal face standard backing coat

profile 13.5/3	m2	6.44	5.0	6.76
profile 19	m2	5.97	5.0	6.27
profile MS20	m2	6.37	5.0	6.69
profile 32	m2	6.64	5.0	6.97
profile 35	m2	7.09	5.0	7.44
profile 38A	m2	6.97	5.0	7.32
profile 40	m2	6.37	5.0	6.69
profile 46	m2	7.81	5.0	8.20
profile 60	m2	8.79	5.0	9.23
profile 100	m2	10.05	5.0	10.55

Cladding, 0.55mm thick, external face HP200, internal face standard backing coat

profile 13.5/3	m2	5.85	5.0	6.14
profile 19	m2	5.43	5.0	5.70
profile MS20	m2	5.80	5.0	6.09
profile 32	m2	6.04	5.0	6.34
profile 35	m2	6.44	5.0	6.76
profile 38A	m2	6.34	5.0	6.66
profile 40	m2	5.80	5.0	6.09
profile 46	m2	6.91	5.0	7.26
profile 60	m2	7.78	5.0	8.17

MATERIAL COSTS

Steel sheeting (cont'd)

	Unit	Material supply (£)	Material waste (%)	Total (£)

Cladding, 0.70mm thick, external face Pvf2, internal face standard backing coat

profile 13.5/3	m2	6.70	5.0	7.04
profile 19	m2	6.23	5.0	6.54
profile MS20	m2	6.64	5.0	6.97
profile 32	m2	6.92	5.0	7.27
profile 35	m2	7.38	5.0	7.75
profile 38A	m2	7.27	5.0	7.63
profile 40	m2	6.64	5.0	6.97
profile 46	m2	7.81	5.0	8.20
profile 60	m2	8.79	5.0	9.23
profile 100	m2	10.05	5.0	10.55

Cladding, 0.70mm thick, external face lining enamel, internal face standard backing coat

profile 13.5/3	m2	6.11	5.0	6.42
profile 19	m2	5.67	5.0	5.95
profile MS20	m2	6.06	5.0	6.36
profile 32	m2	6.31	5.0	6.63
profile 35	m2	7.78	5.0	8.17
profile 38A	m2	6.63	5.0	6.96
profile 40	m2	6.06	5.0	6.36
profile 46	m2	7.04	5.0	7.39
profile 60	m2	7.92	5.0	8.32
profile 100	m2	9.05	5.0	9.50

Colour coated galvanized steel pressed profiled sheets

Cladding, 0.70mm thick, external face HP200, internal face standard backing coat

profile PR8	m2	8.24	5.0	8.65
profile PM13	m2	8.81	5.0	9.25
profile PL19	m2	9.10	5.0	9.56
profile PG22	m2	9.66	5.0	10.14
profile PS47	m2	11.95	5.0	12.55

MATERIAL COSTS

	Unit	Material supply (£)	Material waste (%)	Total (£)

Cladding, 0.55mm thick, external face HP200, internal face standard backing coat

profile PR8	m2	7.30	5.0	7.67
profile PM13	m2	7.80	5.0	8.19
profile PL19	m2	8.06	5.0	8.46
profile PG22	m2	8.55	5.0	8.98
profile PS47	m2	10.57	5.0	11.10

Cladding, 0.70mm thick, external face Pvf2, internal face standard backing coat

profile PR8	m2	9.16	5.0	9.62
profile PM13	m2	9.80	5.0	10.29
profile PL19	m2	10.12	5.0	10.63
profile PG22	m2	10.73	5.0	11.27
profile PS47	m2	13.28	5.0	13.94

Colour coated galvanized steel lining systems

Roll-formed panels 0.40mm thick, external face lining enamel, internal face standard backing coat

profile CL3/900	m2	3.76	5.0	3.95
profile CL3/914	m2	3.71	5.0	3.90
profile CL6/914	m2	3.71	5.0	3.90
profile CL3/960	m2	3.53	5.0	3.71
profile CL3/1000	m2	3.49	5.0	3.66
profile CL3/1016	m2	3.43	5.0	3.60

Roll-formed panels 0.70mm thick, external face lining enamel, internal face standard backing coat

profile 13.5/3	m2	6.11	5.0	6.42
profile 19	m2	5.67	5.0	5.95
profile MS20	m2	6.06	5.0	6.36

MATERIAL COSTS

Galvanised steel (cont'd)	Unit	Material supply (£)	Material waste (%)	Total (£)
Structural lining tray 0.90mm thick, external face lining enamel, internal face standard backing coat				
profile HL70/400	m2	13.50	5.0	14.18
Shadow line tray 0.70mm thick, external finish lining enamel, internal face standard backing coat				
up to 408mm total girth	m	2.74	5.0	2.88
up to 613mm total girth	m	4.12	5.0	4.33
Shadow line tray 0.70mm thick, external face HP200, internal face standard backing coat				
up to 408mm total girth	m	3.19	5.0	3.35
up to 613mm total girth	m	4.78	5.0	5.02

Insulation

Polystyrene (EPS)

profile C19, thickness 62mm	m2	4.66	5.0	4.89
profile R62, thickness 65mm	m2	7.17	5.0	7.53
profile C32, thickness 53mm	m2	7.00	5.0	7.35
profile R38A, thickness 66mm	m2	6.97	5.0	7.32
profile C38A, thickness 50mm	m2	7.12	5.0	7.48
profile R40, thickness 67mm	m2	5.31	5.0	5.58
profile C40, thickness 43mm	m2	5.31	5.0	5.58

Rockfibre 100

profile R32, thickness 67mm	m2	15.21	5.0	15.97
profile C32, thickness 54mm	m2	15.21	5.0	15.97
profile R38A, thickness 68mm	m2	15.59	5.0	16.37
profile C38A, thickness 52mm	m2	15.59	5.0	16.37
profile R40, thickness 69mm	m2	16.26	5.0	17.07
profile C40, thickness 45mm	m2	16.26	5.0	17.07

MATERIAL COSTS

	Unit	Material supply (£)	Material waste (%)	Total (£)

Flashings

Aluminium alloy, plain or stucco

0.70mm thick
100mm girth	m	1.33	5.0	1.40
153mm girth	m	1.81	5.0	1.90
204mm girth	m	2.40	5.0	2.52
245mm girth	m	2.89	5.0	3.03
306mm girth	m	3.60	5.0	3.78
350mm girth	m	4.67	5.0	4.90
408mm girth	m	4.81	5.0	5.05
450mm girth	m	6.01	5.0	6.31
500mm girth	m	6.67	5.0	7.00
612mm girth	m	7.21	5.0	7.57
700mm girth	m	9.33	5.0	9.80
800mm girth	m	10.68	5.0	11.21
900mm girth	m	12.01	5.0	12.61
1000mm girth	m	13.35	5.0	14.02

0.90mm thick
100mm girth	m	1.68	5.0	1.76
153mm girth	m	2.28	5.0	2.39
204mm girth	m	3.03	5.0	3.18
245mm girth	m	3.64	5.0	3.82
306mm girth	m	4.55	5.0	4.78
350mm girth	m	5.89	5.0	6.18
408mm girth	m	6.07	5.0	6.37
450mm girth	m	7.58	5.0	7.96
500mm girth	m	8.42	5.0	8.84
612mm girth	m	9.10	5.0	9.56
700mm girth	m	11.78	5.0	12.37
800mm girth	m	13.47	5.0	14.14
900mm girth	m	15.15	5.0	15.91
1000mm girth	m	16.83	5.0	17.67

Aluminium alloy, plain

1.20mm thick
100mm girth	m	2.18	5.0	2.29
153mm girth	m	2.96	5.0	3.11
204mm girth	m	3.95	5.0	4.15
245mm girth	m	4.74	5.0	4.98
306mm girth	m	5.91	5.0	6.21

MATERIAL COSTS

Aluminium (cont'd)	Unit	Material supply (£)	Material waste (%)	Total (£)
350mm girth	m	7.65	5.0	8.03
408mm girth	m	7.89	5.0	8.28
450mm girth	m	9.85	5.0	10.34
500mm girth	m	10.94	5.0	11.49
612mm girth	m	11.83	5.0	12.42
700mm girth	m	15.32	5.0	16.09
800mm girth	m	17.50	5.0	18.38
900mm girth	m	19.70	5.0	20.69
1000mm girth	m	21.88	5.0	22.97
1.60mm thick				
100mm girth	m	2.87	5.0	3.01
153mm girth	m	3.87	5.0	4.06
204mm girth	m	5.17	5.0	5.43
245mm girth	m	6.21	5.0	6.52
306mm girth	m	7.75	5.0	8.14
350mm girth	m	10.04	5.0	10.54
408mm girth	m	10.33	5.0	10.85
450mm girth	m	12.90	5.0	13.55
500mm girth	m	14.33	5.0	15.05
612mm girth	m	15.50	5.0	16.27
700mm girth	m	20.08	5.0	21.08
800mm girth	m	22.94	5.0	24.09
900mm girth	m	25.81	5.0	27.10
1000mm girth	m	28.68	5.0	30.11

Aluminium alloy, AD80/standard lacquer

0.70mm thick				
100mm girth	m	2.37	5.0	2.49
153mm girth	m	3.20	5.0	3.36
204mm girth	m	4.27	5.0	4.48
245mm girth	m	5.13	5.0	5.39
306mm girth	m	6.42	5.0	6.74
350mm girth	m	8.31	5.0	8.73
408mm girth	m	8.55	5.0	8.98
450mm girth	m	10.68	5.0	11.21
500mm girth	m	11.87	5.0	12.46
612mm girth	m	12.83	5.0	13.47
700mm girth	m	16.61	5.0	17.44
800mm girth	m	18.98	5.0	19.93
900mm girth	m	21.34	5.0	22.41
1000mm girth.	m	23.73	5.0	24.92

MATERIAL COSTS

	Unit	Material supply (£)	Material waste (%)	Total (£)
0.90mm thick				
100mm girth	m	2.94	5.0	3.09
153mm girth	m	3.98	5.0	4.18
204mm girth	m	5.30	5.0	5.56
245mm girth	m	6.37	5.0	6.69
306mm girth	m	7.96	5.0	8.36
350mm girth	m	10.31	5.0	10.83
408mm girth	m	10.62	5.0	11.15
450mm girth	m	13.25	5.0	13.91
500mm girth	m	14.72	5.0	15.46
612mm girth	m	15.92	5.0	16.72
700mm girth	m	20.61	5.0	21.64
800mm girth	m	23.56	5.0	24.74
900mm girth	m	26.50	5.0	27.83
1000mm girth	m	29.45	5.0	30.92
Aluminium alloy, galvanized steel				
0.70mm thick				
100mm girth	m	0.88	5.0	0.92
153mm girth	m	1.52	5.0	1.60
204mm girth	m	2.03	5.0	2.13
245mm girth	m	2.44	5.0	2.56
306mm girth	m	3.04	5.0	3.19
350mm girth	m	4.09	5.0	4.29
408mm girth	m	4.05	5.0	4.25
450mm girth	m	5.26	5.0	5.52
500mm girth	m	5.85	5.0	6.14
612mm girth	m	6.09	5.0	6.39
700mm girth	m	8.19	5.0	8.60
800mm girth	m	9.36	5.0	9.83
900mm girth	m	10.53	5.0	11.06
1000mm girth	m	11.70	5.0	12.29
0.90mm thick				
100mm girth	m	1.10	5.0	1.16
153mm girth	m	1.91	5.0	2.01
204mm girth	m	2.54	5.0	2.67
245mm girth	m	3.06	5.0	3.21
306mm girth	m	3.81	5.0	4.00
350mm girth	m	5.13	5.0	5.39
408mm girth	m	5.09	5.0	5.34
450mm girth	m	6.60	5.0	6.93
500mm girth	m	7.34	5.0	7.71

MATERIAL COSTS

Galvanized steel (cont'd)	Unit	Material supply (£)	Material waste (%)	Total (£)
612mm girth	m	7.63	5.0	8.01
700mm girth	m	10.27	5.0	10.78
800mm girth	m	11.74	5.0	12.33
900mm girth	m	13.21	5.0	13.87
1000mm girth	m	14.67	5.0	15.40
1.20mm thick				
100mm girth	m	1.43	5.0	1.50
153mm girth	m	2.47	5.0	2.59
204mm girth	m	3.30	5.0	3.47
245mm girth	m	3.96	5.0	4.16
306mm girth	m	4.95	5.0	5.20
350mm girth	m	6.66	5.0	6.99
408mm girth	m	6.59	5.0	6.92
450mm girth	m	8.56	5.0	8.99
500mm girth	m	9.50	5.0	9.98
612mm girth	m	9.89	5.0	10.38
700mm girth	m	13.30	5.0	13.97
800mm girth	m	15.20	5.0	15.96
900mm girth	m	17.10	5.0	17.96
1000mm girth	m	19.10	5.0	20.06
1.60mm thick				
100mm girth	m	1.85	5.0	1.94
153mm girth	m	3.20	5.0	3.36
204mm girth	m	4.26	5.0	4.47
245mm girth	m	5.12	5.0	5.38
306mm girth	m	6.40	5.0	6.72
350mm girth	m	8.61	5.0	9.04
408mm girth	m	8.54	5.0	8.97
450mm girth	m	11.08	5.0	11.63
500mm girth	m	12.31	5.0	12.93
612mm girth	m	12.80	5.0	13.44
700mm girth	m	17.22	5.0	18.08
800mm girth	m	19.69	5.0	20.67
900mm girth	m	22.14	5.0	23.25
1000mm girth	m	24.60	5.0	25.83
2.00mm thick				
100mm girth	m	2.31	5.0	2.43
153mm girth	m	4.70	5.0	4.94
204mm girth	m	6.27	5.0	6.58
245mm girth	m	7.53	5.0	7.91
306mm girth	m	9.41	5.0	9.88

MATERIAL COSTS

	Unit	Material supply (£)	Material waste (%)	Total (£)
350mm girth	m	10.76	5.0	11.30
408mm girth	m	12.55	5.0	13.18
450mm girth	m	13.84	5.0	14.53
500mm girth	m	15.37	5.0	16.14
612mm girth	m	18.82	5.0	19.76
700mm girth	m	21.52	5.0	22.60
800mm girth	m	24.59	5.0	25.82
900mm girth	m	27.67	5.0	29.05

Galvanized steel, coloured

0.70mm thick, external face HP200, internal face standard backing coat

	Unit	Material supply (£)	Material waste (%)	Total (£)
100mm girth	m	1.75	5.0	1.84
153mm girth	m	2.80	5.0	2.94
204mm girth	m	3.74	5.0	3.93
245mm girth	m	4.49	5.0	4.71
306mm girth	m	5.61	5.0	5.89
350mm girth	m	8.20	5.0	8.61
408mm girth	m	7.48	5.0	7.85
450mm girth	m	10.55	5.0	11.08
500mm girth	m	11.72	5.0	12.31
612mm girth	m	11.21	5.0	11.77
700mm girth	m	16.41	5.0	17.23
800mm girth	m	18.75	5.0	19.69
900mm girth	m	21.09	5.0	22.14
1000mm girth	m	23.45	5.0	24.62

0.55mm thick, external face HP200, internal face standard backing coat

	Unit	Material supply (£)	Material waste (%)	Total (£)
100mm girth	m	1.58	5.0	1.66
153mm girth	m	2.51	5.0	2.64
204mm girth	m	3.35	5.0	3.52
245mm girth	m	4.02	5.0	4.22
306mm girth	m	5.02	5.0	5.27
350mm girth	m	7.35	5.0	7.72
408mm girth	m	6.70	5.0	7.04
450mm girth	m	9.45	5.0	9.92
500mm girth	m	10.50	5.0	11.03
612mm girth	m	10.04	5.0	10.54
700mm girth	m	14.69	5.0	15.42
800mm girth	m	16.79	5.0	17.63

MATERIAL COSTS

Galvanized steel (cont'd)	Unit	Material supply (£)	Material waste (%)	Total (£)
900mm girth	m	18.89	5.0	19.83
1000mm girth	m	20.99	5.0	22.04
0.70mm thick, external face Pvf2, internal face standard backing coat				
100mm girth	m	1.75	5.0	1.84
153mm girth	m	2.80	5.0	2.94
204mm girth	m	3.74	5.0	3.93
245mm girth	m	4.49	5.0	4.71
306mm girth	m	5.61	5.0	5.89
350mm girth	m	8.20	5.0	8.61
408mm girth	m	7.48	5.0	7.85
450mm girth	m	10.55	5.0	11.08
500mm girth	m	11.72	5.0	12.31
612mm girth	m	11.21	5.0	11.77
700mm girth	m	16.41	5.0	17.23
800mm girth	m	18.75	5.0	19.69
900mm girth	m	21.09	5.0	22.14
1000mm girth	m	23.45	5.0	24.62
0.70mm thick, external face lining enamel, internal face standard backing coat				
100mm girth	m	1.43	5.0	1.50
153mm girth	m	2.28	5.0	2.39
204mm girth	m	3.04	5.0	3.19
245mm girth	m	3.65	5.0	3.83
306mm girth	m	4.57	5.0	4.80
350mm girth	m	6.68	5.0	7.01
408mm girth	m	6.09	5.0	6.39
450mm girth	m	8.59	5.0	9.02
500mm girth	m	9.53	5.0	10.01
612mm girth	m	9.12	5.0	9.58
700mm girth	m	13.36	5.0	14.03
800mm girth	m	15.26	5.0	16.02
900mm girth	m	17.17	5.0	18.03
1000mm girth	m	19.08	5.0	20.03

MATERIAL COSTS

	Unit	Material supply (£)	Material waste (%)	Total (£)
0.40mm thick, external face lining enamel, internal face standard backing coat				
100mm girth	m	0.98	5.0	1.03
153mm girth	m	2.00	5.0	2.10
204mm girth	m	2.67	5.0	2.80
245mm girth	m	3.20	5.0	3.36
306mm girth	m	4.00	5.0	4.20
350mm girth	m	4.57	5.0	4.80
408mm girth	m	5.33	5.0	5.60
450mm girth	m	5.88	5.0	6.17
500mm girth	m	6.53	5.0	6.86
612mm girth	m	7.99	5.0	8.39
700mm girth	m	9.15	5.0	9.61
800mm girth	m	10.45	5.0	10.97
900mm girth	m	11.76	5.0	12.35
1000mm girth	m	13.06	5.0	13.71

MATERIAL COSTS

	Unit	Material supply (£)	Material waste (%)	Total (£)

H62 NATURAL SLATES

Welsh blue/grey slates size

405 x 205mm	100	116.17	2.5	119.07
405 x 255mm	100	152.35	2.5	156.16
405 x 305mm	100	184.55	2.5	189.16
460 x 230mm	100	164.96	2.5	169.08
460 x 255mm	100	184.81	2.5	189.43
460 x 305mm	100	226.92	2.5	232.59
510 x 255mm	100	259.44	2.5	265.93
510 x 305mm	100	295.05	2.5	302.43
560 x 280mm	100	348.61	2.5	357.33
560 x 305mm	100	381.93	2.5	391.48
610 x 305mm	100	514.23	2.5	527.09
ridge tile	m	13.27	2.5	13.60
mitre hip	m	13.95	2.5	14.30

Westmoreland green slates

500-300mm long	ton	1814.25	2.5	1859.61
300-225mm long	ton	1025.00	2.5	1050.63

H61 FIBRE CEMENT SLATES

Duracem

500 x 250mm	100	84.96	5.0	89.21
600 x 300mm	100	113.16	5.0	118.82

Eternit 2000

400 x 240mm	100	62.39	5.0	65.51
500 x 250mm	100	86.72	5.0	91.06
600 x 300mm	100	115.28	5.0	121.04
600 x 600mm	100	230.56	5.0	242.09

Rivendale

500 x 250mm	100	97.07	5.0	101.92
500 x 400mm	100	93.55	5.0	98.23
600 x 300mm	100	129.15	5.0	135.61
600 x 600mm	100	258.29	5.0	271.20

MATERIAL COSTS

	Unit	Material supply (£)	Material waste (%)	Total (£)
Country				
600 x 300mm	100	151.31	2.5	155.09
H63 RECONSTRUCTED STONE SLATES				
Marley Monarch	1000	1314.60	5.0	1380.33
Redland Cambrian	1000	1401.28	5.0	1471.34
H60 CLAY/CONCRETE ROOF TILING				
Marley roof tiles				
Plain	1000	266.50	2.5	273.16
Plain Premium	1000	306.48	2.5	314.14
Feature	1000	391.55	2.5	401.34
Feature Premium	1000	448.95	2.5	460.17
Ludlow Plus	1000	396.68	2.5	406.60
Anglia Plus	1000	472.52	2.5	484.33
Anglia Plus Premium	1000	524.80	2.5	537.92
Ludlow Major	1000	641.65	2.5	657.69
Mendip	1000	727.75	2.5	745.94
Mendip Premium	1000	834.35	2.5	855.21
Supalite Mendip	1000	941.98	2.5	965.53
Double Roman	1000	641.65	2.5	657.69
Double Roman Premium	1000	746.20	2.5	764.86
Modern	1000	732.88	2.5	751.20
Wessex	1000	856.90	2.5	878.32

MATERIAL COSTS

Marley tiles (cont'd)	Unit	Material supply (£)	Material waste (%)	Total (£)
Bold Roll	1000	727.75	2.5	745.94
Bold Roll Premium	1000	834.35	2.5	855.21
Monarch	1000	1283.30	2.5	1315.38
Aluminium nails	1kg	4.95	10.0	5.45
Bonnet hips, valleys and angles	100	163.80	5.0	171.99
Segmental ridge	100	213.15	5.0	223.81
Modern ridge	100	236.25	5.0	248.06
Monoridge - Modern	100	386.40	5.0	405.72
Monoridge - segmental	100	386.40	5.0	405.72
Dentil slips				
Wessex	100	9.45	5.0	9.92
Mendip	100	9.45	5.0	9.92
Bold Roll	100	9.45	5.0	9.92
Verge slips	100	9.45	5.0	9.92
Ventilation ridge terminal	each	28.35	5.0	29.77
Soil vent terminal	each	27.30	5.0	28.67
Gas vent ridge	each	44.10	5.0	46.31
Tile clips				
Modern	100	2.26	5.0	2.37
Wessex	100	2.26	5.0	2.37
Bold Roll	100	2.20	2.5	2.26
Plain	100	5.25	5.0	5.51
Hip irons				
4mm	each	1.26	5.0	1.32
6mm	each	1.37	5.0	1.44
Ventilated dry ridge system				
Modern/Major Batten Section	each	6.04	5.0	6.34

MATERIAL COSTS

	Unit	Material supply (£)	Material waste (%)	Total (£)
Segmental Ridge Union	each	0.70	5.0	0.74
Modern Ridge Union	each	0.70	5.0	0.74
Modern Filler Unit	each	0.30	5.0	0.32
Ludlow Major Filler Unit	each	0.30	5.0	0.32
Mendip/Bold Roll Batten Section	each	6.04	5.0	6.34
Mendip Filler Unit	each	0.30	5.0	0.32
Bold Roll Filler Unit	each	0.30	5.0	0.32
Wessex Filler Unit	each	0.30	5.0	0.32
Anglia Filler Unit	each	0.30	5.0	0.32
Double Roman Filler Unit	each	0.30	5.0	0.32
GVR Adaptor Unit	each	1.98	5.0	2.08
Dry Ridge Setting Out Gauge	each	3.36	5.0	3.53

Eaves ventilation system

Strip Ventilator (1m)	each	1.73	5.0	1.82
Underfelt Support	each	1.25	5.0	1.31
Eaves Vent Duct	each	0.68	5.0	0.71

Profiled eaves fillers

Bold Roll	100	12.07	5.0	12.67
Anglia	100	6.83	5.0	7.17
Mendip	100	1.05	5.0	1.10
Double Roman	100	12.07	5.0	12.67
Wessex	100	6.83	5.0	7.17

Eaves vent clips

Wessex	100	2.63	5.0	2.76
Mendip	100	2.63	5.0	2.76
Modern	100	2.63	5.0	2.76
Ludlow	100	2.63	5.0	2.76
Bold Roll	100	2.63	5.0	2.76
Double Roman	100	2.63	5.0	2.76
Anglia	100	2.63	5.0	2.76

Eaves clips

Modern	100	2.26	5.0	2.37
Mendip	100	2.26	5.0	2.37
Ludlow Major	100	2.26	5.0	2.37
Bold Roll	100	2.26	5.0	2.37
Double Roman	100	2.26	5.0	2.37

MATERIAL COSTS

Marley tiles (cont'd)	Unit	Material supply (£)	Material waste (%)	Total (£)
Verge clips				
Modern	100	19.35	5.0	20.32
Wessex	100	19.35	5.0	20.32
Bold Roll	100	19.63	5.0	20.61
Interlocking dry verge				
LH verge unit	each	1.89	5.0	1.98
RH verge unit	each	1.89	5.0	1.98
LH stop end unit	each	1.89	5.0	1.98
RH stop end unit	each	1.89	5.0	1.98
Segmental ridge end cap	each	1.89	5.0	1.98
Modern ridge end cap	each	1.89	5.0	1.98
LH monoridge end cap	each	3.78	5.0	3.97
RH monoridge end cap	each	3.78	5.0	3.97
Redland tiles				
Clay roof tiling				
Renown	1000	662.94	2.5	679.51
50 Double Roman	1000	662.94	2.5	679.51
Regent	1000	743.84	2.5	762.44
Grovebury	1000	743.84	2.5	762.44
Norfolk Pantile	1000	474.88	2.5	486.75
49	1000	378.23	2.5	387.69
Delta	1000	1088.45	2.5	1115.66
Stonewold MK1	1000	988.63	2.5	1013.35
Stonewold MK2	1000	897.23	2.5	919.66
Richmond	1000	757.50	2.5	776.44
Cambrian	1000	1335.35	2.5	1368.73
Plain	1000	264.76	2.5	271.38
Ornamental	1000	38.87	2.5	39.84
Downland	1000	265.80	2.5	272.45
Rosemary red	1000	349.85	2.5	358.60
Rosemary brindle	1000	439.16	2.5	450.14
Rosemary Cheslyn	1000	559.99	2.5	573.99
Half round ridge	100	216.09	5.0	226.89
Third round hip	100	216.09	5.0	226.89
Universal valley trough	100	441.00	5.0	463.05

MATERIAL COSTS

	Unit	Material supply (£)	Material waste (%)	Total (£)
Dentil slips				
41mm	100	22.05	5.0	23.15
55mm	100	22.05	5.0	23.15
83mm	100	22.05	5.0	23.15
Dryvent ridge system	3m	32.34	5.0	33.96
Angle hip	100	231.53	5.0	243.11
Bonnet hip	100	170.89	5.0	179.43
Valley tiles	100	170.89	5.0	179.43
Gas flue ridge terminal	each	44.13	2.5	45.23
Cloaked verge tile	100	183.86	2.5	188.46
Downland bonnet	100	194.37	2.5	199.23
Universal angle ridge	100	220.63	2.5	226.15
Delta ridge	100	257.41	2.5	263.85
Delta angle hip	100	236.40	2.5	242.31
Dry verge system	5m	39.40	2.5	40.39
Delta flue ridge terminal	each	54.63	2.5	56.00
Shingles				
Random width x 400mm length	pack	30.18	7.5	32.44

MATERIAL COSTS

	Unit	Material supply (£)	Material waste (%)	Total (£)
G32 WOODWOOL SLAB DECKING				
Woodcemair woodwool slabs, unreinforced				
50mm (type 500)				
1800, 2100 and 2400mm lengths	m2	5.58	5.0	5.86
50mm (type 500)				
2700 and 3000mm lengths	m2	5.64	5.0	5.92
75mm (type 750)				
2100mm lengths	m2	7.27	5.0	7.63
75mm (type 750)				
2400 and 2700mm lengths	m2	7.39	5.0	7.76
75mm (type 750)				
3000mm lengths	m2	9.71	5.0	10.20
100mm (type 1000)				
3000, 3800mm lengths	m2	13.37	5.0	14.04
Woodcelip woodwool slabs				
50mm (type 503)				
1800, 2000 and 2100mm lengths	m2	13.15	5.0	13.81
50mm (type 503)				
2400mm lengths	m2	13.80	5.0	14.49
50mm (type 503)				
2700 and 3000mm lengths	m2	14.03	5.0	14.73

MATERIAL COSTS

	Unit	Material supply (£)	Material waste (%)	Total (£)
75mm (type 751) 1800, 2000 and 2400mm lengths	m2	19.34	5.0	20.31
75mm (type 751) 2700 and 3000mm lengths	m2	19.38	5.0	20.35
75mm (type 752) 1800, 2000 and 2400mm lengths	m2	19.23	5.0	20.19
75mm (type 752) 2700 and 3000mm lengths	m2	19.34	5.0	20.31
75mm (type 753) 2400mm lengths	m2	19.18	5.0	20.14
75mm (type 753) 2700 and 3000mm lengths	m2	20.01	5.0	21.01
75mm (type 753) 3300, 3600 and 3900mm lengths	m2	23.47	5.0	24.64
100mm (type 1001) 3000mm lengths	m2	25.27	5.0	26.53
100mm (type 1001) 3300 and 3600mm lengths	m2	26.37	5.0	27.69
100mm (type 1002) 3000mm lengths	m2	24.69	5.0	25.92
100mm (type 1002) 3300 and 3600mm lengths	m2	26.24	5.0	27.55

MATERIAL COSTS

Woodwool slab decking (ont'd)	Unit	Material supply (£)	Material waste (%)	Total (£)
100mm (type 1003)				
3000, 3300 and 3600mm lengths	m2	26.24	5.0	27.55
100mm (type 1003)				
3900 and 4000mm lengths	m2	23.68	5.0	24.86
125mm (type 1252)				
2400, 2700 and 3000mm lengths	m2	26.76	5.0	28.10
Sundries				
Softwood roofing battens				
32 x 19mm	100m	20.16	7.5	21.67
32 x 25mm	100m	25.80	7.5	27.74
38 x 22mm	100m	23.18	7.5	24.92
38 x 25mm	100m	28.78	7.5	30.94
50 x 25mm	100m	37.86	7.5	40.70
Roofing felt				
type 1F reinforced underlay	15m2	13.20	7.5	14.19
H30 FIBRE CEMENT SHEETING				
Corrugated reinforced cement sheeting - Eternit 2000				
profile 3 grey	m2	7.33	5.0	7.70
profile 3 coloured	m2	8.53	5.0	8.96
profile 6 grey	m2	7.51	5.0	7.89
profile 6 coloured	m2	8.65	5.0	9.08
Profile 3 grey fittings				
ridge fittings	nr	5.23	5.0	5.49
eaves filler	nr	5.11	5.0	5.37
eaves closure 75mm	nr	5.11	5.0	5.37
apron flashing	nr	5.84	5.0	6.13

MATERIAL COSTS

	Unit	Material supply (£)	Material waste (%)	Total (£)
Profile 6 grey fittings				
ridge fittings	nr	6.42	5.0	6.74
eaves filler	nr	6.70	5.0	7.04
eaves closure 100mm	nr	7.22	5.0	7.58
eaves bend sheet 1525mm (300mm radius)	nr	24.57	5.0	25.80
apron flashing	nr	7.22	5.0	7.58
Profile 3 coloured fittings				
ridge fittings	nr	6.47	5.0	6.79
eaves filler	nr	5.67	5.0	5.95
eaves closure 75mm	nr	5.95	5.0	6.25
apron flashing	nr	6.51	5.0	6.84
Profile 6 coloured fittings				
ridge fittings	nr	8.45	5.0	8.87
eaves filler	nr	7.77	5.0	8.16
eaves closure 100mm	nr	8.36	5.0	8.78
eaves bendsheet 1525mm (30mm radius)	nr	26.98	5.0	28.33
apron flashing	nr	8.36	5.0	8.78

H41 TRANSLUCENT SHEETING

Corrugated glass fibre reinforced translucent sheeting, 1.3mm thick

75mm profile	m2	6.91	5.0	7.26
150mm profile	m2	10.25	5.0	10.76

H76 FIBRE BITUMEN SHEETING

Nuralite

FX	m2	13.92	5.0	14.62
Nutec FX	m2	14.48	5.0	15.20

MATERIAL COSTS

Fibre bitumen sheeting (cont'd)	Unit	Material supply (£)	Material waste (%)	Total (£)
Beaded cover flashings, preformed, girth				
100mm	m	1.28	5.0	1.34
150mm	m	1.67	5.0	1.75
200mm	m	2.21	5.0	2.32
250mm	m	2.72	5.0	2.86
300mm	m	3.36	5.0	3.53
Ridge trays, preformed, length				
250mm	nr	1.06	5.0	1.11
350mm	nr	1.25	5.0	1.31
450mm	nr	1.49	5.0	1.56
Intermediate trays, preformed, length				
250mm	nr	1.23	5.0	1.29
350mm	nr	1.49	5.0	1.56
450mm	nr	1.63	5.0	1.71
Catchment trays, preformed, length				
250mm	nr	1.25	5.0	1.31
350mm	nr	1.63	5.0	1.71
450mm	nr	1.63	5.0	1.71
Catchment closure pieces, preformed, girth				
250mm	nr	1.80	5.0	1.89
350mm	nr	0.91	5.0	0.96
450mm	nr	1.05	5.0	1.10
Soakers, preformed, girth				
150mm	nr	0.56	5.0	0.59
175mm	nr	0.59	5.0	0.62
216mm	nr	0.72	5.0	0.76
250mm	nr	0.82	5.0	0.86
300mm	nr	0.95	5.0	1.00
350mm	nr	1.10	5.0	1.16
432mm	nr	1.66	5.0	1.74

MATERIAL COSTS

	Unit	Material supply (£)	Material waste (%)	Total (£)
Linings to concrete gutters, preformed, girth				
450mm	m	5.74	5.0	6.03
490mm	m	5.74	5.0	6.03
500mm	m	5.74	5.0	6.03
519mm	m	5.74	5.0	6.03
H71 LEAD SHEET COVERINGS				
Sheet lead to BS1178	t	1060.50	1.0	1071.11
H73 COPPER SHEET COVERINGS				
Sheet copper to BS2870				
0.55mm thick	0.1t	367.50	5.0	385.88
0.7mm thick	0.1t	368.55	5.0	386.98
H74 ZINC SHEET COVERINGS				
Sheet zinc to BS849				
0.65mm thick	0.1t	173.25	5.0	181.91
0.8mm thick	0.1t	165.90	5.0	174.20
H72 ALUMINIUM SHEET COVERINGS				
Sheet aluminium to BS1470				
0.6mm thick				
150mm wide	8m	11.14	5.0	11.70
200mm wide	8m	20.23	5.0	21.24
375mm wide	8m	26.33	5.0	27.65
450mm wide	8m	30.15	5.0	31.66
600mm wide	8m	38.20	5.0	40.11
900mm wide	8m	55.99	5.0	58.79
0.8mm thick				
150mm wide	8m	14.31	5.0	15.03

MATERIAL COSTS

Aluminium sheeting (cont'd)	Unit	Material supply (£)	Material waste (%)	Total (£)
225mm wide	8m	25.98	5.0	27.28
300mm wide	8m	33.83	5.0	35.52
450mm wide	8m	38.74	5.0	40.68
600mm wide	8m	49.07	5.0	51.52
900mm wide	8m	71.92	5.0	75.52

J41 BUILT-UP FELT ROOF COVERINGS

Built-up roofing to BS747

Felt type 1B

14kg/10m2	10m x 1m roll	nr	9.40	5.0	9.87

Felt type 1B

18kg/10m2	10m x 1m roll	nr	12..04	5.0	12.64

Felt type 1B

25kg/10m2	10m x 1m roll	nr	17.28	5.0	18.14

Felt type 1E

38kg/10m2	10m x 1m roll	nr	26.66	5.0	27.99

Felt type 3B

18kg/10m2	20m x 1m roll	nr	27.26	5.0	28.62

Felt type 3E

28kg/10m2	10m x 1m roll	nr	20.42	5.0	21.44

Felt type 3G

28kg/10m2	10m x 1m roll	nr	24.21	5.0	25.42

Felt type 5U

29kg/10m2	10m x 1m roll	nr	59.39	5.0	62.36

MATERIAL COSTS

	Unit	Material supply (£)	Material waste (%)	Total (£)
Felt type 5B				
34kg/10m2 10m x 1m roll	nr	49.85	5.0	52.34
Felt type 5E				
38kg/10m2 10m x 1m roll	nr	57.35	5.0	60.22
Felt type Elastomeric				
40kg/20m2 20m x 1m roll	nr	62.93	5.0	66.08
Felt type Elastomeric				
32kg/10m2 10m x 1m roll	nr	43.52	5.0	45.70
Primo	25l	43.85	10.0	48.24
Adhesive	25l	35.07	10.0	38.58
Felt type Euroroof Elastomeric				
G.32	10m2	28.79	5.0	30.23
P.56	10m2	53.83	5.0	56.52

Foamglas

T2 Slabs

	Unit	Material supply (£)	Material waste (%)	Total (£)
40mm thick	m2	9.60	5.0	10.08
50mm thick	m2	11.57	5.0	12.15
60mm thick	m2	13.58	5.0	14.26
70mm thick	m2	15.68	5.0	16.46
80mm thick	m2	17.85	5.0	18.74
90mm thick	m2	19.98	5.0	20.98
100mm thick	m2	22.17	5.0	23.28
110mm thick	m2	24.32	5.0	25.54
120mm thick	m2	26.44	5.0	27.76
130mm thick	m2	28.58	5.0	30.01

T4 slabs

	Unit	Material supply (£)	Material waste (%)	Total (£)
40mm thick	m2	9.62	5.0	10.10
50mm thick	m2	11.63	5.0	12.21
60mm thick	m2	13.73	5.0	14.42
70mm thick	m2	15.95	5.0	16.75

MATERIAL COSTS

Foamglas (cont'd)	Unit	Material supply (£)	Material waste (%)	Total (£)
80mm thick	m2	18.24	5.0	19.15
90mm thick	m2	20.51	5.0	21.54
100mm thick	m2	23.76	5.0	24.95
110mm thick	m2	25.04	5.0	26.29
120mm thick	m2	27.30	5.0	28.67
130mm thick	m2	29.58	5.0	31.06
140mm thick	m2	31.06	5.0	32.61
150mm thick	m2	33.30	5.0	34.96
S3 slabs				
40mm thick	m2	10.86	5.0	11.40
50mm thick	m2	13.11	5.0	13.77
60mm thick	m2	15.50	5.0	16.27
80mm thick	m2	20.66	5.0	21.69
100mm thick	m2	25.66	5.0	26.94

MATERIAL COSTS

	Unit	Material supply (£)	Material waste (%)	Total (£)

L11 ROOFLIGHTS

'Coxdome Mark 1' rooflight

Single skin clear, diffused or tinted PVC-U domed rooflight

600mm diameter	nr	43.46	2.5	44.55
900mm diameter	nr	62.93	2.5	64.50
1200mm diameter	nr	79.49	2.5	81.48
1800mm diameter	nr	203.46	2.5	208.55

Single skin clear or diffused wire laminate PVC-U domed roof light

600mm diameter	nr	52.58	2.5	53.89
900mm diameter	nr	75.54	2.5	77.43
1200mm diameter	nr	96.04	2.5	98.44

Single skin clear or diffused polycarbonate domed rooflight

600mm diameter	nr	61.60	2.5	63.14
900mm diameter	nr	87.28	2.5	89.46
1200mm diameter	nr	112.60	2.5	115.41
1800mm diameter	nr	288.59	2.5	295.80

Double skin domed rooflight with clear PVC-U standard inner skin and clear, diffused or tinted PVC-U outer skin

600mm diameter	nr	76.72	2.5	78.64
900mm diameter	nr	115.21	2.5	118.09
1200mm diameter	nr	149.60	2.5	153.34
1800mm diameter	nr	383.09	2.5	392.67

MATERIAL COSTS

Rooflights (cont'd)	Unit	Material supply (£)	Material waste (%)	Total (£)
Double skin domed rooflight with clear PVC-U standard inner skin and clear or diffused wire laminate PVC-U outer skin				
600mm diameter	nr	92.66	2.5	94.98
900mm diameter	nr	138.27	2.5	141.73
1200mm diameter	nr	180.76	2.5	185.28
Double skin domed rooflight with clear PVC-U standard inner skin and clear or diffused polycarbonate outer skin				
600mm diameter	nr	108.65	2.5	111.37
900mm diameter	nr	161.34	2.5	165.37
1200mm diameter	nr	211.92	2.5	217.22
1800mm diameter	nr	542.74	2.5	556.31
'Coxdome Mark 2' rooflight				
Single skin clear, diffused or tinted PVC-U domed rooflight				
619 x 619mm	nr	44.74	2.5	45.86
772 x 772mm	nr	52.22	2.5	53.53
924 x 924mm	nr	59.76	2.5	61.25
1076 x 1076mm	nr	59.91	2.5	61.41
1229 x 1229mm	nr	144.68	2.5	148.30
Single skin clear, diffused or tinted PVC-U pyramidal rooflight				
619 x 619mm	nr	80.10	2.5	82.10
772 x 772mm	nr	87.23	2.5	89.41
924 x 924mm	nr	94.30	2.5	96.66
1076 x 1076mm	nr	96.45	2.5	98.86
1229 x 1229mm	nr	177.94	2.5	182.39

MATERIAL COSTS

	Unit	Material supply (£)	Material waste (%)	Total (£)
Single skin clear or diffused wire laminate PVC-U domed rooflight				
619 x 619mm	nr	64.08	2.5	65.68
772 x 772mm	nr	75.75	2.5	77.64
924 x 924mm	nr	86.66	2.5	88.83
1076 x 1076mm	nr	86.92	2.5	89.09
Single skin clear or diffused polycarbonate domed rooflight				
619 x 619mm	nr	76.11	2.5	78.01
772 x 772mm	nr	88.61	2.5	90.83
924 x 924mm	nr	101.63	2.5	104.17
1076 x 1076mm	nr	101.89	2.5	104.44
1229 x 1229mm	nr	246.00	2.5	252.15
Single skin clear or diffused polycarbonate pyramidal rooflight				
619 x 619mm	nr	136.22	2.5	139.63
772 x 772mm	nr	147.34	2.5	151.02
924 x 924mm	nr	160.31	2.5	164.32
1076 x 1076mm	nr	164.00	2.5	168.10
1229 x 1229mm	nr	302.38	2.5	309.94
Double skin domed rooflight with clear PVC-U standard inner skin and clear, diffused or tinted acrylic outer skin				
619 x 619mm	nr	78.72	2.5	80.69
772 x 772mm	nr	93.53	2.5	95.87
924 x 924mm	nr	108.39	2.5	111.10
1076 x 1076mm	nr	109.88	2.5	112.63
1229 x 1229mm	nr	279.67	2.5	286.66

MATERIAL COSTS

Rooflights (cont'd)	Unit	Material supply (£)	Material waste (%)	Total (£)

Double skin pyramidal rooflight with PVC-U standard inner skin and clear, diffused or tinted PVC-U outer skin

619 x 619mm	nr	117.62	2.5	120.56
772 x 772mm	nr	132.12	2.5	135.42
924 x 924mm	nr	146.68	2.5	150.35
1076 x 1076mm	nr	151.75	2.5	155.54
1229 x 1229mm	nr	314.78	2.5	322.65

Double skin domed rooflight with clear PVC-U standard inner skin and clear or diffused wire laminate uPVC outer skin

619 x 619mm	nr	114.19	2.5	117.04
772 x 772mm	nr	135.66	2.5	139.05
924 x 924mm	nr	157.18	2.5	161.11
1076 x 1076mm	nr	159.34	2.5	163.32

Double skin domed rooflight with clear PVC-U standard inner skin and clear or diffused polycarbonate outer skin

619 x 619mm	nr	133.81	2.5	137.16
772 x 772mm	nr	158.47	2.5	162.43
924 x 924mm	nr	184.30	2.5	188.91
1076 x 1076mm	nr	186.81	2.5	191.48
1229 x 1229mm	nr	475.45	2.5	487.34

Double skin pyramidal rooflight with clear PVC-U standard inner skin and clear or diffused acrylic outer skin

619 x 619mm	nr	141.14	2.5	144.67
772 x 772mm	nr	157.85	2.5	161.80
924 x 924mm	nr	176.04	2.5	180.44
1076 x 1076mm	nr	182.09	2.5	186.64
1229 x 1229mm	nr	377.71	2.5	387.15

MATERIAL COSTS

	Unit	Material supply (£)	Material waste (%)	Total (£)

'Coxdome Mark 3' rooflight

Single skin clear, diffused or tinted acrylic domed rooflight

600 x 600mm	nr	35.11	2.5	35.99
900 x 600mm	nr	61.29	2.5	62.82
900 x 900mm	nr	67.86	2.5	69.56
1200 x 900mm	nr	72.01	2.5	73.81
1200 x 1200mm	nr	100.45	2.5	102.96
1800 x 1200mm	nr	165.79	2.5	169.93

Single skin clear, diffused or tinted acrylic pyramidal rooflight

600 x 600mm	nr	38.64	2.5	39.61
900 x 600mm	nr	67.39	2.5	69.07
900 x 900mm	nr	74.62	2.5	76.49
1200 x 900mm	nr	87.94	2.5	90.14
1200 x 1200mm	nr	110.49	2.5	113.25
1800 x 1200mm	nr	182.40	2.5	186.96

Single skin clear or diffused wire laminate PVC-U domed rooflight

600 x 600mm	nr	41.00	2.5	42.03
900 x 600mm	nr	85.08	2.5	87.21
900 x 900mm	nr	93.69	2.5	96.03
1200 x 900mm	nr	104.91	2.5	107.53
1200 x 1200mm	nr	124.23	2.5	127.34
1800 x 1200mm	nr	184.70	2.5	189.32

Single skin clear or diffused PVC-U domed rooflight

600 x 600mm	nr	32.95	2.5	33.77
900 x 600mm	nr	68.06	2.5	69.76
900 x 900mm	nr	68.47	2.5	70.18
1200 x 900mm	nr	82.82	2.5	84.89
1200 x 1200mm	nr	85.95	2.5	88.10
1800 x 1200mm	nr	145.75	2.5	149.39

MATERIAL COSTS

Rooflights (cont'd)	Unit	Material supply (£)	Material waste (%)	Total (£)
Single skin clear or diffused PVC-U pyramidal rooflight				
600 x 600mm	nr	36.23	2.5	37.14
900 x 600mm	nr	74.93	2.5	76.80
900 x 900mm	nr	75.29	2.5	77.17
1200 x 900mm	nr	91.17	2.5	93.45
1200 x 1200mm	nr	94.56	2.5	96.92
1800 x 1200mm	nr	160.31	2.5	164.32
Single skin polycarbonate domed rooflight				
600 x 600mm	nr	59.30	2.5	60.78
900 x 600mm	nr	122.59	2.5	125.65
900 x 900mm	nr	123.26	2.5	126.34
1200 x 900mm	nr	148.98	2.5	152.70
1200 x 1200mm	nr	154.72	2.5	158.59
1800 x 1200mm	nr	262.35	2.5	268.91
Single skin polycarbonate pyramidal rooflight				
600 x 600mm	nr	65.24	2.5	66.87
900 x 600mm	nr	196.34	2.5	201.25
900 x 900mm	nr	135.50	2.5	138.89
1200 x 900mm	nr	164.05	2.5	168.15
1200 x 1200mm	nr	170.15	2.5	174.40
1800 x 1200mm	nr	288.59	2.5	295.80
Double skin domed rooflight with clear PVC-U standard inner skin and clear, diffused or tinted acrylic outer skin				
600 x 600mm	nr	71.85	2.5	73.65
900 x 600mm	nr	106.34	2.5	109.00
900 x 900mm	nr	133.97	2.5	137.32
1200 x 900mm	nr	152.98	2.5	156.80
1200 x 1200mm	nr	186.81	2.5	191.48
1800 x 1200mm	nr	285.15	2.5	292.28

MATERIAL COSTS

	Unit	Material supply (£)	Material waste (%)	Total (£)
Double skin pyramidal rooflight with clear PVC-U standard inner skin and clear, diffused or tinted acrylic outer skin				
600 x 600mm	nr	82.61	2.5	84.68
900 x 600mm	nr	122.33	2.5	125.39
900 x 900mm	nr	154.06	2.5	157.91
1200 x 900mm	nr	175.89	2.5	180.29
1200 x 1200mm	nr	214.79	2.5	220.16
1800 x 1200mm	nr	317.70	2.5	325.64
Double skin domed rooflight with clear PVC-U standard inner skin and clear or diffused wire laminate PVC-U outer skin				
600 x 600mm	nr	83.23	2.5	85.31
900 x 600mm	nr	145.24	2.5	148.87
900 x 900mm	nr	175.99	2.5	180.39
1200 x 900mm	nr	176.86	2.5	181.28
1200 x 1200mm	nr	225.14	2.5	230.77
1800 x 1200mm	nr	324.87	2.5	332.99
Double skin domed rooflight with clear PVC-U standard inner skin and clear or diffused PVC-U outer skin				
600 x 600mm	nr	56.84	2.5	58.26
900 x 600mm	nr	111.83	2.5	114.63
900 x 900mm	nr	113.42	2.5	116.26
1200 x 900mm	nr	127.46	2.5	130.65
1200 x 1200mm	nr	138.27	2.5	141.73
1800 x 1200mm	nr	258.76	2.5	265.23

MATERIAL COSTS

Rooflights (cont'd)

	Unit	Material supply (£)	Material waste (%)	Total (£)
Double skin pyramidal rooflight with clear PVC-U standard inner skin and clear or diffused PVC-U outer skin				
600 x 600mm	nr	65.29	2.5	66.92
900 x 600mm	nr	128.64	2.5	131.86
900 x 900mm	nr	130.38	2.5	133.64
1200 x 900mm	nr	146.57	2.5	150.23
1200 x 1200mm	nr	159.03	2.5	163.01
1800 x 1200mm	nr	297.56	2.5	305.00
Double skin domed rooflight with clear PVC-U standard inner skin and clear, diffused or tinted polycarbonate outer skin				
600 x 600mm	nr	102.35	2.5	104.91
900 x 600mm	nr	166.31	2.5	170.47
900 x 900mm	nr	204.13	2.5	209.23
1200 x 900mm	nr	217.50	2.5	222.94
1200 x 1200mm	nr	276.90	2.5	283.82
1800 x 1200mm	nr	369.26	2.5	378.49
Double skin pyramidal rooflight with clear PVC-U standard inner skin and clear, diffused or tinted polycarbonate outer skin				
600 x 600mm	nr	117.67	2.5	120.61
900 x 600mm	nr	191.21	2.5	195.99
900 x 900mm	nr	234.72	2.5	240.59
1200 x 900mm	nr	250.72	2.5	256.99
1200 x 1200mm	nr	318.42	2.5	326.38
1800 x 1200mm	nr	431.68	2.5	442.47

MATERIAL COSTS

	Unit	Material supply (£)	Material waste (%)	Total (£)

'Coxdome Mark 4' rooflight

Single skin clear, diffused or tinted PVC-U domed rooflight

600 x 600mm	nr	88.51	2.5	90.72
900 x 600mm	nr	150.21	2.5	153.97
900 x 900mm	nr	158.26	2.5	162.22
1200 x 900mm	nr	194.95	2.5	199.82
1200 x 1200mm	nr	207.36	2.5	212.54
1800 x 1200mm	nr	309.65	2.5	317.39

Single skin clear, diffused or tinted PVC-U pyramidal rooflight

600 x 600mm	nr	91.79	2.5	94.08
900 x 600mm	nr	157.08	2.5	161.01
900 x 900mm	nr	165.08	2.5	169.21
1200 x 900mm	nr	203.31	2.5	208.39
1200 x 1200mm	nr	215.97	2.5	221.37
1800 x 1200mm	nr	324.21	2.5	332.32

Single skin clear or diffused wire laminate PVC-U domed rooflight

600 x 600mm	nr	96.56	2.5	98.97
900 x 600mm	nr	167.23	2.5	171.41
900 x 900mm	nr	183.47	2.5	188.06
1200 x 900mm	nr	216.92	2.5	222.34
1200 x 1200mm	nr	237.03	2.5	242.96
1800 x 1200mm	nr	362.49	2.5	371.55

Single skin clear or diffused acrylic domed rooflight

600 x 600mm	nr	89.64	2.5	91.88
900 x 600mm	nr	143.45	2.5	147.04
900 x 900mm	nr	157.65	2.5	161.59
1200 x 900mm	nr	191.93	2.5	196.73
1200 x 1200mm	nr	213.25	2.5	218.58
1800 x 1200mm	nr	343.58	2.5	352.17

MATERIAL COSTS

Rooflights (cont'd)	Unit	Material supply (£)	Material waste (%)	Total (£)
Single skin clear or diffused acrylic pyramidal rooflight				
600 x 600mm	nr	93.28	2.5	95.61
900 x 600mm	nr	149.55	2.5	153.29
900 x 900mm	nr	164.41	2.5	168.52
1200 x 900mm	nr	199.98	2.5	204.98
1200 x 1200mm	nr	223.30	2.5	228.88
1800 x 1200mm	nr	360.19	2.5	369.19
Double skin domed rooflight with clear PVC-U standard inner skin and clear, diffused or tinted PVC-U outer skin				
600 x 600mm	nr	102.14	2.5	104.69
900 x 600mm	nr	193.98	2.5	198.83
900 x 900mm	nr	203.21	2.5	208.29
1200 x 900mm	nr	238.11	2.5	244.06
1200 x 1200mm	nr	251.07	2.5	257.35
1800 x 1200mm	nr	436.55	2.5	447.46
Double skin pyramidal rooflight with clear PVC-U standard inner skin and clear, diffused or tinted PVC-U outer skin				
600 x 600mm	nr	110.60	2.5	113.37
900 x 600mm	nr	210.79	2.5	216.06
900 x 900mm	nr	220.17	2.5	225.67
1200 x 900mm	nr	257.43	2.5	263.87
1200 x 1200mm	nr	271.83	2.5	278.63
1800 x 1200mm	nr	475.34	2.5	487.22

MATERIAL COSTS

	Unit	Material supply (£)	Material waste (%)	Total (£)
Double skin domed rooflight with clear PVC-U standard inner skin and clear or diffused wire laminate PVC-U outer skin				
600 x 600mm	nr	138.78	2.5	142.25
900 x 600mm	nr	227.40	2.5	233.09
900 x 900mm	nr	265.78	2.5	272.42
1200 x 900mm	nr	288.90	2.5	296.12
1200 x 1200mm	nr	337.94	2.5	346.39
1800 x 1200mm	nr	488.77	2.5	500.99
Double skin domed rooflight with clear PVC-U standard inner skin and clear or diffused acrylic outer skin				
600 x 600mm	nr	127.41	2.5	130.60
900 x 600mm	nr	188.50	2.5	193.21
900 x 900mm	nr	223.76	2.5	229.35
1200 x 900mm	nr	265.01	2.5	271.64
1200 x 1200mm	nr	299.61	2.5	307.10
1800 x 1200mm	nr	462.94	2.5	474.51
Double skin pyramidal rooflight with clear PVC-U standard inner skin and clear or diffused acrylic outer skin				
600 x 600mm	nr	138.17	2.5	141.62
900 x 600mm	nr	204.49	2.5	209.60
900 x 900mm	nr	243.85	2.5	249.95
1200 x 900mm	nr	287.92	2.5	295.12
1200 x 1200mm	nr	327.59	2.5	335.78
1800 x 1200mm	nr	505.73	2.5	518.37

MATERIAL COSTS

	Unit	Material supply (£)	Material waste (%)	Total (£)

'Coxdome Mark 4' rooflight

Single skin clear, diffused or tinted PVC-U domed rooflight

600 x 600mm	nr	145.81	2.5	149.46
900 x 600mm	nr	202.69	2.5	207.76
900 x 900mm	nr	219.76	2.5	225.25
1200 x 900mm	nr	264.14	2.5	270.74
1200 x 1200mm	nr	189.01	2.5	193.74
1800 x 1200mm	nr	400.88	2.5	410.90

Single skin clear, diffused or tinted PVC-U pyramidal rooflight

600 x 600mm	nr	149.09	2.5	152.82
900 x 600mm	nr	209.56	2.5	214.80
900 x 900mm	nr	226.68	2.5	232.35
1200 x 900mm	nr	272.48	2.5	279.29
1200 x 1200mm	nr	197.21	2.5	202.14
1800 x 1200mm	nr	415.43	2.5	425.82

Single skin clear or diffused wire laminate PVC-U domed rooflight

600 x 600mm	nr	153.85	2.5	157.70
900 x 600mm	nr	219.71	2.5	225.20
900 x 900mm	nr	244.97	2.5	251.09
1200 x 900mm	nr	305.71	2.5	313.35
1200 x 1200mm	nr	339.22	2.5	347.70
1800 x 1200mm	nr	439.62	2.5	450.61

Single skin clear or diffused acrylic domed rooflight

600 x 600mm	nr	147.96	2.5	151.66
900 x 600mm	nr	195.93	2.5	200.83
900 x 900mm	nr	219.15	2.5	224.63
1200 x 900mm	nr	279.98	2.5	286.98
1200 x 1200mm	nr	208.13	2.5	213.33
1800 x 1200mm	nr	420.71	2.5	431.23

MATERIAL COSTS

	Unit	Material supply (£)	Material waste (%)	Total (£)
Single skin clear or diffused acrylic pyramidal rooflight				
600 x 600mm	nr	151.60	2.5	155.39
900 x 600mm	nr	202.03	2.5	207.08
900 x 900mm	nr	225.91	2.5	231.56
1200 x 900mm	nr	287.92	2.5	295.12
1200 x 1200mm	nr	218.17	2.5	223.62
1800 x 1200mm	nr	437.32	2.5	448.25
Double skin domed rooflight with clear PVC-U standard inner skin and clear, diffused or tinted PVC-U outer skin				
600 x 600mm	nr	169.69	2.5	173.93
900 x 600mm	nr	246.46	2.5	252.62
900 x 900mm	nr	264.71	2.5	271.33
1200 x 900mm	nr	308.78	2.5	316.50
1200 x 1200mm	nr	355.57	2.5	364.46
1800 x 1200mm	nr	552.48	2.5	566.29
Double skin pyramidal rooflight with clear PVC-U standard inner skin and clear, diffused or tinted PVC-U outer skin				
600 x 600mm	nr	178.15	2.5	182.60
900 x 600mm	nr	263.27	2.5	269.85
900 x 900mm	nr	281.77	2.5	288.81
1200 x 900mm	nr	327.80	2.5	336.00
1200 x 1200mm	nr	376.33	2.5	385.74
1800 x 1200mm	nr	581.02	2.5	595.55

MATERIAL COSTS

Rooflights (cont'd)

	Unit	Material supply (£)	Material waste (%)	Total (£)
Double skin domed rooflight with clear PVC-U standard inner skin and clear or diffused wire laminate PVC-U outer skin				
600 x 600mm	nr	196.08	2.5	200.98
900 x 600mm	nr	279.88	2.5	286.88
900 x 900mm	nr	327.28	2.5	335.46
1200 x 900mm	nr	358.23	2.5	367.19
1200 x 1200mm	nr	421.69	2.5	432.23
1800 x 1200mm	nr	620.79	2.5	636.31
Double skin domed rooflight with clear PVC-U standard inner skin and clear or diffused acrylic outer skin				
600 x 600mm	nr	184.70	2.5	189.32
900 x 600mm	nr	240.98	2.5	247.00
900 x 900mm	nr	285.26	2.5	292.39
1200 x 900mm	nr	334.34	2.5	342.70
1200 x 1200mm	nr	383.35	2.5	392.93
1800 x 1200mm	nr	523.67	2.5	536.76
Double skin pyramidal rooflight with clear PVC-U standard inner skin and clear or diffused acrylic outer skin				
600 x 600mm	nr	195.47	2.5	200.36
900 x 600mm	nr	256.97	2.5	263.39
900 x 900mm	nr	304.73	2.5	312.35
1200 x 900mm	nr	357.11	2.5	366.04
1200 x 1200mm	nr	411.33	2.5	421.61
1800 x 1200mm	nr	566.47	2.5	580.63

MATERIAL COSTS

	Unit	Material supply (£)	Material waste (%)	Total (£)

'Coxdome Mark 4' rooflight

Single skin clear, diffused or tinted PVC-U domed rooflight

600 x 600mm	nr	170.00	2.5	174.25
900 x 600mm	nr	222.17	2.5	227.72
900 x 900mm	nr	262.49	2.5	269.05
1200 x 900mm	nr	282.80	2.5	289.87
1200 x 1200mm	nr	296.12	2.5	303.52
1800 x 1200mm	nr	399.80	2.5	409.80

Single skin clear, diffused or tinted PVC-U pyramidal rooflight

600 x 600mm	nr	173.28	2.5	177.61
900 x 600mm	nr	228.93	2.5	234.65
900 x 900mm	nr	270.40	2.5	277.16
1200 x 900mm	nr	291.15	2.5	298.43
1200 x 1200mm	nr	304.73	2.5	312.35
1800 x 1200mm	nr	414.36	2.5	424.72

Single skin clear or diffused wire laminate PVC-U domed rooflight

600 x 600mm	nr	182.14	2.5	186.69
900 x 600mm	nr	239.18	2.5	245.16
900 x 900mm	nr	277.42	2.5	284.36
1200 x 900mm	nr	305.33	2.5	312.96
1200 x 1200mm	nr	334.41	2.5	342.77
1800 x 1200mm	nr	438.75	2.5	449.72

Single skin clear or diffused acrylic domed rooflight

600 x 600mm	nr	170.04	2.5	174.29
900 x 600mm	nr	222.17	2.5	227.72
900 x 900mm	nr	262.45	2.5	269.01
1200 x 900mm	nr	282.80	2.5	289.87
1200 x 1200mm	nr	296.12	2.5	303.52
1800 x 1200mm	nr	399.80	2.5	409.80

MATERIAL COSTS

Rooflights (cont'd)	Unit	Material supply (£)	Material waste (%)	Total (£)
Single skin clear or diffused acrylic pyramidal rooflight				
600 x 600mm	nr	173.53	2.5	177.87
900 x 600mm	nr	228.27	2.5	233.98
900 x 900mm	nr	269.22	2.5	275.95
1200 x 900mm	nr	244.62	2.5	250.74
1200 x 1200mm	nr	306.17	2.5	313.82
1800 x 1200mm	nr	415.79	2.5	426.18
Double skin domed rooflight with clear PVC-U standard inner skin and clear, diffused or tinted PVC-U outer skin				
600 x 600mm	nr	193.88	2.5	198.73
900 x 600mm	nr	266.14	2.5	272.79
900 x 900mm	nr	307.40	2.5	315.09
1200 x 900mm	nr	327.44	2.5	335.63
1200 x 1200mm	nr	348.45	2.5	357.16
1800 x 1200mm	nr	512.89	2.5	525.71
Double skin pyramidal rooflight with clear PVC-U standard inner skin and clear, diffused or tinted PVC-U outer skin				
600 x 600mm	nr	202.34	2.5	207.40
900 x 600mm	nr	282.75	2.5	289.82
900 x 900mm	nr	324.36	2.5	332.47
1200 x 900mm	nr	346.55	2.5	355.21
1200 x 1200mm	nr	369.20	2.5	378.43
1800 x 1200mm	nr	551.60	2.5	565.39

MATERIAL COSTS

	Unit	Material supply (£)	Material waste (%)	Total (£)
Double skin domed rooflight with clear PVC-U standard inner skin and clear or diffused wire laminate outer skin				
600 x 600mm	nr	220.27	2.5	225.78
900 x 600mm	nr	299.35	2.5	306.83
900 x 900mm	nr	369.97	2.5	379.22
1200 x 900mm	nr	376.74	2.5	386.16
1200 x 1200mm	nr	435.32	2.5	446.20
1800 x 1200mm	nr	578.92	2.5	593.39
Double domed rooflight with clear PVC-U standard inner skin and clear or diffused acrylic outer skin				
600 x 600mm	nr	208.90	2.5	214.12
900 x 600mm	nr	260.45	2.5	266.96
900 x 900mm	nr	327.95	2.5	336.15
1200 x 900mm	nr	342.71	2.5	351.28
1200 x 1200mm	nr	396.98	2.5	406.90
1800 x 1200mm	nr	539.20	2.5	552.68
Double skin pyramidal rooflight with clear PVC-U standard inner skin and clear or diffused acrylic outer skin				
600 x 600mm	nr	202.59	2.5	207.65
900 x 600mm	nr	281.93	2.5	288.98
900 x 900mm	nr	327.49	2.5	335.68
1200 x 900mm	nr	350.35	2.5	359.11
1200 x 1200mm	nr	376.38	2.5	385.79
1800 x 1200mm	nr	555.60	2.5	569.49

MATERIAL COSTS

	Unit	Material supply (£)	Material waste (%)	Total (£)

'Coxdome 2000' rooflight fixed to PVC-U adaptor plugged and screwed to flat builder's curb as distributed by Coxdome Ltd

Double skin domed rooflight with clear PVC-U inner skin and clear, diffused or tinted PVC-U outer skin

600 x 600mm	nr	157.03	2.5	160.96
900 x 600mm	nr	212.48	2.5	217.79
900 x 900mm	nr	223.09	2.5	228.67
1200 x 900mm	nr	264.55	2.5	271.16
1200 x 1200mm	nr	306.32	2.5	313.98
1800 x 1200mm	nr	422.50	2.5	433.06

Double skin pyramidal rooflight with clear PVC-U inner skin and clear, diffused or tinted PVC-U outer skin

600 x 600mm	nr	173.69	2.5	178.03
900 x 600mm	nr	236.93	2.5	242.85
900 x 900mm	nr	248.46	2.5	254.67
1200 x 900mm	nr	296.02	2.5	303.42
1200 x 1200mm	nr	342.35	2.5	350.91

Double skin domed rooflight with clear acrylic inner skin and clear or diffused acrylic outer skin

600 x 600mm	nr	162.92	2.5	166.99
900 x 600mm	nr	203.26	2.5	208.34
900 x 900mm	nr	237.08	2.5	243.01
1200 x 900mm	nr	271.93	2.5	278.73
1200 x 1200mm	nr	334.87	2.5	343.24
1800 x 1200mm	nr	471.60	2.5	483.39

MATERIAL COSTS

	Unit	Material supply (£)	Material waste (%)	Total (£)
Double skin pyramidal rooflight with clear acrylic inner skin and clear or diffused acrylic outer skin				
600 x 600mm	nr	180.55	2.5	185.06
900 x 600mm	nr	226.32	2.5	231.98
900 x 900mm	nr	264.55	2.5	271.16
1200 x 900mm	nr	304.48	2.5	312.09
1200 x 1200mm	nr	375.46	2.5	384.85
Double skin domed rooflight with clear wired laminate PVC-U inner skin and clear or diffused PVC-U outer skin				
600 x 600mm	nr	171.79	2.5	176.08
900 x 600mm	nr	264.96	2.5	271.58
900 x 900mm	nr	274.85	2.5	281.72
1200 x 900mm	nr	300.43	2.5	307.94
1200 x 1200mm	nr	354.24	2.5	363.10
1800 x 1200mm	nr	503.17	2.5	515.75
Double skin domed rooflight with clear PVC-U inner skin and clear or diffused polycarbonate outer skin				
600 x 600mm	nr	220.07	2.5	225.57
900 x 600mm	nr	351.11	2.5	359.89
900 x 900mm	nr	363.26	2.5	372.34
1200 x 900mm	nr	428.96	2.5	439.68
1200 x 1200mm	nr	510.45	2.5	523.21
1800 x 1200mm	nr	735.18	2.5	753.56

MATERIAL COSTS

	Unit	Material supply (£)	Material waste (%)	Total (£)
'Coxdome 2000' rooflight				
Double skin domed rooflight with clear PVC-U inner skin and clear, diffused or tinted PVC-U outer skin				
600 x 600mm	nr	226.22	2.5	231.88
900 x 600mm	nr	291.15	2.5	298.43
900 x 900mm	nr	313.65	2.5	321.49
1200 x 900mm	nr	360.80	2.5	369.82
1200 x 1200mm	nr	312.88	2.5	320.70
1800 x 1200mm	nr	551.60	2.5	565.39
Double skin pyramidal rooflight with clear PVC-U inner skin and clear, diffused or tinted PVC-U outer skin				
600 x 600mm	nr	241.85	2.5	247.90
900 x 600mm	nr	315.60	2.5	323.49
900 x 900mm	nr	340.04	2.5	348.54
1200 x 900mm	nr	392.27	2.5	402.08
1200 x 1200mm	nr	451.67	2.5	462.96
Double skin domed rooflight with clear acrylic inner skin and clear or diffused acrylic outer skin				
600 x 600mm	nr	231.09	2.5	236.87
900 x 600mm	nr	281.93	2.5	288.98
900 x 900mm	nr	327.64	2.5	335.83
1200 x 900mm	nr	368.18	2.5	377.38
1200 x 1200mm	nr	447.93	2.5	459.13
1800 x 1200mm	nr	600.70	2.5	615.72

MATERIAL COSTS

	Unit	Material supply (£)	Material waste (%)	Total (£)
Double skin pyramidal rooflight with clear acrylic inner skin and clear or diffused acrylic outer skin				
600 x 600mm	nr	248.72	2.5	254.94
900 x 600mm	nr	304.99	2.5	312.61
900 x 900mm	nr	355.11	2.5	363.99
1200 x 900mm	nr	400.72	2.5	410.74
1200 x 1200mm	nr	504.56	2.5	517.17
Double skin domed rooflight with clear wired laminate PVC-U inner skin and clear or diffused PVC-U outer skin				
600 x 600mm	nr	239.95	2.5	245.95
900 x 600mm	nr	343.63	2.5	352.22
900 x 900mm	nr	365.41	2.5	374.55
1200 x 900mm	nr	396.68	2.5	406.60
1200 x 1200mm	nr	463.30	2.5	474.88
1800 x 1200mm	nr	632.27	2.5	648.08
Double skin domed rooflight with clear PVC-U inner skin and clear or diffused polycarbonate outer skin				
600 x 600mm	nr	288.23	2.5	295.44
900 x 600mm	nr	429.78	2.5	440.52
900 x 900mm	nr	453.82	2.5	465.17
1200 x 900mm	nr	525.21	2.5	538.34
1200 x 1200mm	nr	619.51	2.5	635.00
1800 x 1200mm	nr	864.28	2.5	885.89

MATERIAL COSTS

	Unit	Material supply (£)	Material waste (%)	Total (£)

'VELUX' ROOF WINDOWS

'Velux' roof windows type GGL pivoted; fixing brackets screwed to sloping rafters; aluminium clad externally

Double glazed 2 x 3mm panels clear float glass

GGL1; 780 x 980mm	nr	134.89	2.5	138.26
GGL2; 780 x 1400mm	nr	162.83	2.5	166.90
GGL3; 940 x 1600mm	nr	194.15	2.5	199.00
GGL4; 1140 x 1180mm	nr	184.27	2.5	188.88
GGL5; 700 x 1180mm	nr	141.88	2.5	145.43
GGL6; 550 x 980mm	nr	117.55	2.5	120.49
GGL7; 1340 x 980mm	nr	186.68	2.5	191.35
GGL8; 1340 x 1400mm	nr	220.41	2.5	225.92
GGL9; 550 x 700mm	nr	143.61	2.5	147.20

Double glazed; anti-sun glass and 3mm clear float glass

GGL1; 780 x 980mm	nr	158.23	2.5	162.19
GGL2; 780 x 1400mm	nr	190.49	2.5	195.25
GGL3; 940 x 1600mm	nr	231.25	2.5	237.03
GGL4; 1140 x 1180mm	nr	218.36	2.5	223.82
GGL5; 700 x 1180mm	nr	165.54	2.5	169.68
GGL6; 550 x 980mm	nr	137.50	2.5	140.94
GGL7; 1340 x 980mm	nr	219.97	2.5	225.47
GGL8; 1340 x 1400mm	nr	263.04	2.5	269.62
GGL9; 550 x 700mm	nr	120.23	2.5	123.24

MATERIAL COSTS

	Unit	Material supply (£)	Material waste (%)	Total (£)

'Velux' roof windows type GHL, top lining fixing brackets

Double glazed; 2 x 3mm clear float glass

GHL1; 780 x 980mm	nr	114.80	2.5	117.67
GHL2; 780 x 1400mm	nr	138.58	2.5	142.04
GHL4; 1140 x 1180mm	nr	165.23	2.5	169.36
GHL5; 700 x 1180mm	nr	156.82	2.5	160.74
GHL7; 1340 x 980mm	nr	120.74	2.5	123.76
GHL8; 1340 x 1400mm	nr	100.04	2.5	102.54

Double glazed; anti-sun glass and 3mm clear float glass

GHL1; 780 x 980mm	nr	172.67	2.5	176.99
GHL2; 780 x 1400mm	nr	202.11	2.5	207.16
GHL4; 1140 x 1180mm	nr	227.60	2.5	233.29
GHL5; 700 x 1180mm	nr	180.93	2.5	185.45
GHL7; 1340 x 980mm	nr	227.06	2.5	232.74
GHL8; 1340 x 1400mm	nr	267.08	2.5	273.76

'Velux' roof windows type VFE vertical window element; aluminium clad externally

VFE1/60; 780 x 980mm	nr	104.13	2.5	106.73
VFE/1 95; 780 x 980mm	nr	116.68	2.5	119.60
VFE/2 60; 780 x 140mm	nr	104.13	2.5	106.73
VFE/2 95; 780 x 140mm	nr	116.68	2.5	119.60
VFE/3 60; 940 x 1600mm	nr	115.70	2.5	118.59
VFE3/95; 940 x 1600mm	nr	128.90	2.5	132.12
VFE4/60; 1140 x 1180mm	nr	128.08	2.5	131.28
VFE/4 95; 1140 x 1180mm	nr	142.92	2.5	146.49

MATERIAL COSTS

	Unit	Material supply (£)	Material waste (%)	Total (£)
'Velux' aluminium preformed flashings and linings				
Type U for tiles and pantiles to suit window size				
780 x 980mm	nr	24.68	2.5	25.30
780 x 1400mm	nr	27.30	2.5	27.98
940 x 1600mm	nr	30.57	2.5	31.33
1140 x 1180mm	nr	29.91	2.5	30.66
700 x 1180mm	nr	24.68	2.5	25.30
550 x 980mm	nr	21.74	2.5	22.28
1340 x 980mm	nr	31.21	2.5	31.99
1340 x 1400mm	nr	33.67	2.5	34.51
550 x 700mm	nr	20.10	2.5	20.60
Type H for profiled roofing to suit window size				
780 x 980mm	nr	28.51	2.5	29.22
780 x 1400mm	nr	31.33	2.5	32.11
940 x 1600mm	nr	35.99	2.5	36.89
1140 x 1180mm	nr	35.58	2.5	36.47
700 x 1180mm	nr	29.52	2.5	30.26
550 x 980mm	nr	25.47	2.5	26.11
1340 x 980mm	nr	37.20	2.5	38.13
1340 x 1400mm	nr	39.83	2.5	40.83
550 x 700mm	nr	23.85	2.5	24.45
Type L for thin slates to suit window size				
780 x 980mm	nr	21.67	2.5	22.21
780 x 1400mm	nr	24.82	2.5	25.44
940 x 1600mm	nr	27.96	2.5	28.66
1140 x 1180mm	nr	26.11	2.5	26.76
700 x 1180mm	nr	22.78	2.5	23.35
550 x 980mm	nr	20.00	2.5	20.50
1340 x 980mm	nr	26.29	2.5	26.95
1340 x 1400mm	nr	29.25	2.5	29.98
550 x 700mm	nr	16.85	2.5	17.27

MATERIAL COSTS

	Unit	Material supply (£)	Material waste (%)	Total (£)
Type HF for combined GGL and VFE vertical window element 600 or 800mm high				
780 x 980mm	nr	33.93	2.5	34.78
780 x 1400mm	nr	36.35	2.5	37.26
940 x 1600mm	nr	39.09	2.5	40.07
1140 x 1180mm	nr	38.45	2.5	39.41
1340 x 980mm	nr	38.77	2.5	39.74
1340 x 1400mm	nr	41.03	2.5	42.06
Type LF for combined GGL and VFE vertical window element 600 or 800mm high				
780 x 980mm	nr	44.74	2.5	45.86
780 x 1400mm	nr	44.74	2.5	45.86
940 x 1600mm	nr	47.82	2.5	49.02
1140 x 1180mm	nr	48.30	2.5	49.51
1340 x 980mm	nr	50.77	2.5	52.04
1340 x 1400mm	nr	50.56	2.5	51.82
Type HBH 100 for coupled windows horizontally with 100mm frame gap; for profiled roofing materials				
780 x 980mm	nr	60.78	2.5	62.30
780 x 1400mm	nr	64.47	2.5	66.08
940 x 1600mm	nr	71.39	2.5	73.17
1140 x 1180mm	nr	71.34	2.5	73.12
700 x 1180mm	nr	61.29	2.5	62.82
550 x 980mm	nr	54.63	2.5	56.00
1340 x 980mm	nr	76.36	2.5	78.27
1340 x 1400mm	nr	79.95	2.5	81.95
550 x 700mm	nr	52.22	2.5	53.53

MATERIAL COSTS

	Unit	Material supply (£)	Material waste (%)	Total (£)
'Velux' interior lining systems				
Type LSA E 30/60 or E 40/50				
780 x 980mm	nr	45.10	2.5	46.23
780 x 1400mm	nr	49.76	2.5	51.00
940 x 1600mm	nr	53.48	2.5	54.82
1140 x 1180mm	nr	52.37	2.5	53.68
700 x 1180mm	nr	46.40	2.5	47.56
550 x 980mm	nr	42.12	2.5	43.17
1340 x 980mm	nr	42.30	2.5	43.36
1340 x 1400mm	nr	56.84	2.5	58.26
550 x 700mm	nr	39.32	2.5	40.30
'Velux' blinds and awnings				
Roller blinds self-coloured for windows size				
780 x 980mm	nr	26.12	2.5	26.77
780 x 1400mm	nr	31.71	2.5	32.50
940 x 1600mm	nr	38.81	2.5	39.78
1140 x 1180mm	nr	37.69	2.5	38.63
700 x 1180mm	nr	27.98	2.5	28.68
550 x 980mm	nr	21.65	2.5	22.19
1340 x 980mm	nr	38.44	2.5	39.40
1340 x 1400mm	nr	45.90	2.5	47.05
550 x 700mm	nr	18.65	2.5	19.12
Venetian blinds standard cords for windows size				
780 x 980mm	nr	60.34	2.5	61.85
780 x 1400mm	nr	72.42	2.5	74.23
940 x 1600mm	nr	86.89	2.5	89.06
1140 x 1180mm	nr	81.72	2.5	83.76
700 x 1180mm	nr	61.73	2.5	63.27
550 x 980mm	nr	53.79	2.5	55.13
1340 x 980mm	nr	90.34	2.5	92.60
1340 x 1400mm	nr	109.31	2.5	112.04
550 x 700mm	nr	46.89	2.5	48.06

G STRUCTURAL CARCASSING METAL/TIMBER

G32 Woodwool decking

WOODWOOL SLAB DECKING

	Unit	Labour hours	Net labour (£)	Net material (£)	O'heads /profit (£)	Total (£)

G32 WOODWOOL SLAB DECKING

Woodcemair unreinforced woodwool slabs (type 5B) in standard lengths, fixed to timber joists, thickness 50mm (type 500)

1800mm lengths	m2	0.68	4.76	5.70	1.57	12.03
2100mm lengths	m2	0.68	4.76	6.30	1.66	12.72
2400mm lengths	m2	0.68	4.76	6.30	1.66	12.72
2700mm lengths	m2	0.68	4.76	6.37	1.67	12.80
3000mm lengths	m2	0.68	4.76	6.37	1.67	12.80

Woodcemair unreinforced woodwool slabs (type 5B) in standard lengths, fixed to timber joists, thickness 75mm (type 750)

2100mm lengths	m2	0.78	5.46	9.77	2.28	17.51
2400mm lengths	m2	0.78	5.46	10.05	2.33	17.84
2700mm lengths	m2	0.78	5.46	10.05	2.33	17.84
3000mm lengths	m2	0.78	5.46	10.20	2.35	18.01

Woodcemair unreinforced woodwool slabs (type 5B) in standard lengths, fixed to timber joists, thickness 100mm (type 1000)

3000mm lengths	m2	0.86	6.02	14.04	3.01	23.07
3300mm lengths	m2	0.86	6.02	14.04	3.01	23.07
3800mm lengths	m2	0.86	6.02	14.04	3.01	23.07

Woodcelip reinforced woodwool slabs, in standard lengths, fixed to timber joists, thickness 50mm (type 503)

1800mm lengths	m2	0.90	6.30	13.81	3.02	23.13
2000mm lengths	m2	0.90	6.30	13.81	3.02	23.13
2100mm lengths	m2	0.90	6.30	13.81	3.02	23.13
2400mm lengths	m2	0.90	6.30	14.49	3.12	23.91
2700mm lengths	m2	0.90	6.30	14.73	3.15	24.18
3000mm lengths	m2	0.90	6.30	14.73	3.15	24.18

RATES FOR MEASURED WORK

Woodcelip slabs (cont'd)	Unit	Labour hours	Net labour (£)	Net material (£)	O'heads /profit (£)	Total (£)
Woodcelip reinforced woodwool slabs, in standard lengths, fixed to timber joists, thickness 75mm (type 751)						
1800mm lengths	m2	0.98	6.86	20.31	4.08	31.25
2000mm lengths	m2	0.98	6.86	20.31	4.08	31.25
2400mm lengths	m2	0.98	6.86	20.31	4.08	31.25
2700mm lengths	m2	0.98	6.86	20.35	4.08	31.29
3000mm lengths	m2	0.98	6.86	20.35	4.08	31.29
Woodcelip reinforced woodwool slabs, in standard lengths, fixed to timber joists, thickness 75mm (type 752)						
1800mm lengths	m2	0.98	6.86	20.19	4.06	31.11
2000mm lengths	m2	0.98	6.86	20.19	4.06	31.11
2400mm lengths	m2	0.98	6.86	20.19	4.06	31.11
2700mm lengths	m2	0.98	6.86	20.31	4.08	31.25
3000mm lengths	m2	0.98	6.86	20.31	4.08	31.25
Woodcelip reinforced woodwool slabs, in standard lengths, fixed to timber joists, thickness 75mm (type 753)						
2400mm lengths	m2	0.98	6.86	20.14	4.05	31.05
2700mm lengths	m2	0.98	6.86	21.01	4.18	32.05
3000mm lengths	m2	0.98	6.86	21.01	4.18	32.05
3300mm lengths	m2	0.98	6.86	24.64	4.72	36.22
3600mm lengths	m2	0.98	6.86	24.64	4.72	36.22
3900mm lengths	m2	0.98	6.86	24.64	4.72	36.22
Woodcelip reinforced woodwool slabs, in standard lengths, fixed to timber joists, thickness 100mm (type 1001)						
3000mm lengths	m2	1.12	7.84	26.53	5.16	39.53
3300mm lengths	m2	1.12	7.84	27.69	5.33	40.86
3600mm lengths	m2	1.12	7.84	27.69	5.33	40.86

WOODWOOL SLAB DECKING

	Unit	Labour hours	Net labour (£)	Net material (£)	O'heads /profit (£)	Total (£)
Woodcelip reinforced woodwool slabs, in standard lengths, fixed to timber joists, thickness 100mm (type 1002)						
3000mm lengths	m2	1.12	7.84	25.92	5.06	38.82
3300mm lengths	m2	1.12	7.84	27.55	5.31	40.70
3600mm lengths	m2	1.12	7.84	27.55	5.31	40.70
Woodcelip reinforced woodwool slabs, in standard lengths, fixed to timber joists, thickness 100mm (type 1003)						
3000mm lengths	m2	1.12	7.84	24.86	4.90	37.60
3300mm lengths	m2	1.12	7.84	24.86	4.90	37.60
3600mm lengths	m2	1.12	7.84	24.86	4.90	37.60
3900mm lengths	m2	1.12	7.84	24.86	4.90	37.60
4000mm lengths	m2	1.12	7.84	24.86	4.90	37.60
Woodcelip reinforced woodwool slabs, in standard lengths, fixed to timber joists, thickness 125mm (type 1252)						
2400mm lengths	m2	1.15	8.05	28.10	5.42	41.57
2700mm lengths	m2	1.15	8.05	28.10	5.42	41.57
3000mm lengths	m2	1.15	8.05	28.10	5.42	41.57

H CLADDING/COVERING

H30 Fibre cement sheet cladding/covering
H31 Metal profiled flat sheet cladding/covering
H32 Plastics profiled sheet cladding/cladding
H41 Translucent sheeting
H60 Clay/concrete roof tiling
    Underfelt and battens
H61 Fibre cement slating
H62 Natural slating
H64 Timber shingling
H71 Lead sheet covering
H72 Aluminium sheet covering
H73 Copper sheet covering
H74 Zinc sheet covering
H76 Nuralite

FIBRE CEMENT CLADDING

	Unit	Labour hours	Net labour (£)	Net material (£)	O'heads /profit (£)	Total (£)

H30 FIBRE CEMENT CLADDING

Corrugated reinforced cement sheeting, lapped one corrugation at sides and 150mm at ends, fixed with screws and washers to timber purlins

profile 3 grey sheets	m2	0.75	5.25	8.09	2.00	15.34
profile 3 coloured sheets	m2	0.75	5.25	9.41	2.20	16.86
profile 6 grey sheets	m2	0.70	4.90	8.28	1.98	15.16
profile 6 coloured sheets	m2	0.70	4.90	9.53	2.16	16.59

Corrugated reinforced cement sheeting, lapped one corrugation at sides and 150mm at ends, fixed with hook bolts and washers to steel purlins

profile 3 grey sheets	m2	0.85	5.95	8.09	2.11	16.15
profile 3 coloured sheets	m2	0.85	5.95	9.41	2.30	17.66
profile 6 grey sheets	m2	0.80	5.60	8.28	2.08	15.96
profile 6 coloured sheets	m2	0.80	5.60	9.53	2.27	17.40

Fittings to profile 3 sheets

ridge fitting	nr	0.35	2.45	5.76	1.23	9.44
eaves filler	nr	0.20	1.40	5.64	1.06	8.10
eaves closure 75mm	nr	0.20	1.40	5.64	1.06	8.10
apron flashing	nr	0.25	1.75	6.44	1.23	9.42

Fittings to profile 6 sheets

ridge fitting	nr	0.35	2.45	7.08	1.43	10.96
eaves closure 100mm	nr	0.20	1.40	7.96	1.40	10.76
eaves bend sheet 1525mm (300mm radius)	nr	0.35	2.45	27.09	4.43	33.97
apron flashing	nr	0.25	1.75	7.96	1.46	11.17

RATES FOR MEASURED WORK

	Unit	Labour hours	Net labour (£)	Net material (£)	O'heads /profit (£)	Total (£)

H31 METAL PROFILED SHEETING (PRECISION METAL FORMING LTD)

Aluminium alloy roll-formed profiled sheets

Wall cladding, fixed to steelwork with self-tapping screws, 0.70mm thick, plain or stucco

profile 13.5/3	m2	0.40	2.80	4.94	1.16	8.90
profile 19	m2	0.42	2.94	4.59	1.13	8.66
profile MS20	m2	0.45	3.15	4.89	1.21	9.25
profile 32	m2	0.47	3.29	5.09	1.26	9.64
profile 35	m2	0.48	3.36	5.44	1.32	10.12
profile 38A	m2	0.50	3.50	5.46	1.34	10.30
profile 40	m2	0.50	3.50	4.66	1.22	9.38
profile 46	m2	0.50	3.50	5.18	1.30	9.98
profile 60	m2	0.56	3.92	5.82	1.46	11.20

Wall cladding, fixed to steelwork with self-tapping screws, 0.90mm thick, plain or stucco

profile 13.5/3	m2	0.40	2.80	6.23	1.35	10.38
profile 19	m2	0.42	2.94	5.79	1.31	10.04
profile MS20	m2	0.45	3.15	5.88	1.35	10.38
profile 32	m2	0.45	3.15	6.43	1.44	11.02
profile 35	m2	0.47	3.29	6.86	1.52	11.67
profile 38A	m2	0.48	3.36	6.75	1.52	11.63
profile 40	m2	0.50	3.50	5.88	1.41	10.79
profile 46	m2	0.50	3.50	6.86	1.55	11.91
profile 60	m2	0.56	3.92	7.72	1.75	13.39
profile 100	m2	0.60	4.20	8.40	1.89	14.49

Wall cladding, fixed to steelwork with self-tapping screws, 1.20mm thick, plain

profile 13.5/3	m2	0.40	2.80	8.10	1.63	12.53
profile 19	m2	0.42	2.94	7.52	1.57	12.03
profile MS20	m2	0.45	3.15	7.63	1.62	12.40
profile 32	m2	0.45	3.15	8.36	1.73	13.24
profile 35	m2	0.47	3.29	8.90	1.83	14.02

METAL PROFILED SHEETING

	Unit	Labour hours	Net labour (£)	Net material (£)	O'heads /profit (£)	Total (£)
profile 38A	m2	0.48	3.36	8.78	1.82	13.96
profile 40	m2	0.50	3.50	7.63	1.67	12.80
profile 46	m2	0.50	3.50	8.90	1.86	14.26
profile 60	m2	0.56	3.92	10.02	2.09	16.03
profile 100	m2	0.60	4.20	11.46	2.35	18.01

Wall cladding, fixed to steelwork with self-tapping screws, 0.70mm thick, AD80/standard backing coat

	Unit	Labour hours	Net labour (£)	Net material (£)	O'heads /profit (£)	Total (£)
profile 13.5/3	m2	0.40	2.80	8.78	1.74	13.32
profile 19	m2	0.42	2.94	8.15	1.66	12.75
profile MS20	m2	0.45	3.15	8.69	1.78	13.62
profile 32	m2	0.45	3.15	9.06	1.83	14.04
profile 35	m2	0.47	3.29	9.66	1.94	14.89
profile 38A	m2	0.48	3.36	9.51	1.93	14.80
profile 40	m2	0.50	3.50	8.69	1.83	14.02
profile 46	m2	0.50	3.50	9.66	1.97	15.13
profile 60	m2	0.56	3.92	10.98	2.23	17.13
profile 100	m2	0.60	4.20	12.42	2.49	19.11

Roof cladding, fixed to steelwork with self-tapping screws, 0.70mm thick, plain or stucco

	Unit	Labour hours	Net labour (£)	Net material (£)	O'heads /profit (£)	Total (£)
profile 13.5/3	m2	0.35	2.45	4.94	1.11	8.50
profile 19	m2	0.37	2.59	4.59	1.08	8.26
profile MS20	m2	0.39	2.73	4.89	1.14	8.76
profile 32	m2	0.39	2.73	5.09	1.17	8.99
profile 35	m2	0.40	2.80	5.44	1.24	9.48
profile 38A	m2	0.42	2.94	5.34	1.24	9.52
profile 40	m2	0.44	3.08	4.89	1.20	9.17
profile 46	m2	0.44	3.08	5.44	1.28	9.80
profile 60	m2	0.56	3.92	6.11	1.50	11.53

Roof cladding, fixed to steelwork with self-tapping screws, 0.90mm thick, plain or stucco

	Unit	Labour hours	Net labour (£)	Net material (£)	O'heads /profit (£)	Total (£)
profile 13.5/3	m2	0.35	2.45	6.23	1.30	9.98
profile 19	m2	0.37	2.59	5.79	1.26	9.64
profile MS20	m2	0.39	2.73	6.17	1.33	10.23
profile 32	m2	0.39	2.73	6.43	1.37	10.53
profile 35	m2	0.40	2.80	6.86	1.45	11.11

RATES FOR MEASURED WORK

Aluminium sheeting (cont'd)	Unit	Labour hours	Net labour (£)	Net material (£)	O'heads /profit (£)	Total (£)
profile 38A	m2	0.42	2.94	6.75	1.45	11.14
profile 40	m2	0.44	3.08	6.17	1.39	10.64
profile 46	m2	0.44	3.08	6.86	1.49	11.43
profile 60	m2	0.52	3.64	7.72	1.70	13.06
profile 100	m2	0.60	4.20	8.82	1.95	14.97

Roof cladding, fixed to steelwork with self-tapping screws, 1.20mm thick, plain

profile 13.5/3	m2	0.33	2.31	8.10	1.56	11.97
profile 19	m2	0.37	2.59	7.52	1.52	11.63
profile MS20	m2	0.39	2.73	8.01	1.61	12.35
profile 32	m2	0.39	2.73	8.36	1.66	12.75
profile 35	m2	0.40	2.80	8.90	1.75	13.45
profile 38A	m2	0.42	2.94	8.78	1.76	13.48
profile 40	m2	0.44	3.08	8.01	1.66	12.75
profile 46	m2	0.44	3.08	8.90	1.80	13.78
profile 60	m2	0.52	3.64	10.02	2.05	15.71
profile 100	m2	0.60	4.20	11.46	2.35	18.01

Roof cladding, fixed to steelwork with self-tapping screws, 0.70mm thick, AD80/standard backing coat

profile 13.5/3	m2	0.35	2.45	8.78	1.68	12.91
profile 19	m2	0.37	2.59	8.15	1.61	12.35
profile MS20	m2	0.39	2.73	8.69	1.71	13.13
profile 32	m2	0.39	2.73	9.06	1.77	13.56
profile 35	m2	0.40	2.80	9.66	1.87	14.33
profile 38A	m2	0.42	2.94	9.51	1.87	14.32
profile 40	m2	0.44	3.08	8.69	1.77	13.54
profile 46	m2	0.44	3.08	9.66	1.91	14.65
profile 60	m2	0.52	3.64	10.98	2.19	16.81
profile 100	m2	0.60	4.20	12.42	2.49	19.11

Roof cladding, fixed to steelwork with self-tapping screws, 0.90mm thick, AD80/standard lacquer

profile 13.5/3	m2	0.40	2.80	10.89	2.05	15.74
profile 19	m2	0.42	2.94	10.12	1.96	15.02
profile MS20	m2	0.39	2.73	10.79	2.03	15.55
profile 32	m2	0.39	2.73	11.25·	2.10	16.08

METAL PROFILED SHEETING

	Unit	Labour hours	Net labour (£)	Net material (£)	O'heads /profit (£)	Total (£)
profile 35	m2	0.40	2.80	11.99	2.22	17.01
profile 38A	m2	0.42	2.94	11.81	2.21	16.96
profile 40	m2	0.44	3.08	10.79	2.08	15.95
profile 46	m2	0.44	3.08	11.99	2.26	17.33
profile 60	m2	0.52	3.64	13.49	2.57	19.70
profile 100	m2	0.56	3.92	15.42	2.90	22.24

Aluminium alloy pressed sheets

Wall cladding, fixed to steelwork with self-tapping screws, 0.70mm thick, plain or stucco

	Unit	Labour hours	Net labour (£)	Net material (£)	O'heads /profit (£)	Total (£)
profile PR8	m2	0.42	2.94	5.55	1.27	9.76
profile PM13	m2	0.43	3.01	5.94	1.34	10.29
profile PL19	m2	0.44	3.08	6.14	1.38	10.60
profile PG22	m2	0.45	3.15	6.52	1.45	11.12
profile PS47	m2	0.56	3.92	8.06	1.80	13.78

Wall cladding, fixed to steelwork with self-tapping screws, 0.90mm thick, plain or stucco

	Unit	Labour hours	Net labour (£)	Net material (£)	O'heads /profit (£)	Total (£)
profile PR8	m2	0.42	2.94	7.01	1.49	11.44
profile PM13	m2	0.43	3.01	7.50	1.58	12.09
profile PL19	m2	0.44	3.08	7.75	1.62	12.45
profile PG22	m2	0.45	3.15	8.22	1.71	13.08
profile PS47	m2	0.56	3.92	10.17	2.11	16.20

Wall cladding, fixed to steelwork with self-tapping screws, 1.20mm thick, plain

	Unit	Labour hours	Net labour (£)	Net material (£)	O'heads /profit (£)	Total (£)
profile PR8	m2	0.42	2.94	9.11	1.81	13.86
profile PL19	m2	0.44	3.08	10.08	1.97	15.13
profile PG22	m2	0.45	3.15	10.68	2.07	15.90
profile PS47	m2	0.56	3.92	13.22	2.57	19.71

Wall cladding, fixed to steelwork with self-tapping screws, 0.70mm thick, AD80/standard backing coat

	Unit	Labour hours	Net labour (£)	Net material (£)	O'heads /profit (£)	Total (£)
profile PR8	m2	0.42	2.94	9.89	1.92	14.75
profile PM13	m2	0.43	3.01	10.57	2.04	15.62

RATES FOR MEASURED WORK

Aluminium sheeting (cont'd)	Unit	Labour hours	Net labour (£)	Net material (£)	O'heads /profit (£)	Total (£)
profile PL19	m2	0.44	3.08	10.93	2.10	16.11
profile PG22	m2	0.45	3.15	11.59	2.21	16.95
profile PS47	m2	0.56	3.92	14.34	2.74	21.00

Wall cladding, fixed to steelwork with self-tapping screws, 0.90mm thick, AD80/standard lacquer

profile PR8	m2	0.37	2.59	12.27	2.23	17.09
profile PM13	m2	0.39	2.73	13.13	2.38	18.24
profile PL19	m2	0.41	2.87	13.57	2.47	18.91
profile PG22	m2	0.43	3.01	14.39	2.61	20.01
profile PS47	m2	0.44	3.08	17.80	3.13	24.01

Roof cladding, fixed to steelwork with self-tapping screws, 0.70mm thick, plain or stucco

profile PR8	m2	0.39	2.73	5.55	1.24	9.52
profile PM13	m2	0.39	2.73	5.94	1.30	9.97
profile PL19	m2	0.41	2.87	6.14	1.35	10.36
profile PG22	m2	0.43	3.01	6.52	1.43	10.96
profile PS47	m2	0.44	3.08	8.06	1.67	12.81

Roof cladding, fixed to steelwork with self-tapping screws, 0.90mm thick, plain or stucco

profile PR8	m2	0.37	2.59	7.01	1.44	11.04
profile PM13	m2	0.39	2.73	7.50	1.53	11.76
profile PL19	m2	0.41	2.87	7.75	1.59	12.21
profile PG22	m2	0.43	3.01	8.22	1.68	12.91
profile PS47	m2	0.44	3.08	10.17	1.99	15.24

Roof cladding, fixed to steelwork with self-tapping screws, 1.20mm thick, plain

profile PR8	m2	0.37	2.59	9.11	1.75	13.45
profile PL19	m2	0.41	2.87	10.08	1.94	14.89
profile PG22	m2	0.43	3.01	10.68	2.05	15.74
profile PS47	m2	0.44	3.08	13.22	2.44	18.74

METAL PROFILED SHEETING

	Unit	Labour hours	Net labour (£)	Net material (£)	O'heads /profit (£)	Total (£)
Roof cladding, fixed to steelwork with self-tapping screws, 0.70mm thick, AD80/standard backing coat						
profile PR8	m2	0.37	2.59	9.89	1.87	14.35
profile PM13	m2	0.39	2.73	10.57	1.99	15.29
profile PL19	m2	0.41	2.87	10.93	2.07	15.87
profile PG22	m2	0.43	3.01	11.59	2.19	16.79
profile PS47	m2	0.44	3.08	14.34	2.61	20.03
Roof cladding, fixed to steelwork with self-tapping screws, 0.90mm thick, AD80/standard lacquer						
profile PR8	m2	0.37	2.59	12.27	2.23	17.09
profile PM13	m2	0.39	2.73	13.13	2.38	18.24
profile PL19	m2	0.41	2.87	13.57	2.47	18.91
profile PG22	m2	0.43	3.01	14.39	2.61	20.01
profile PS47	m2	0.44	3.08	17.80	3.13	24.01

Galvanized steel roll-formed profiled sheets

Wall cladding, fixed to steelwork with self-tapping screws, 0.70mm thick

	Unit					
profile 13.5/3	m2	0.40	2.80	4.86	1.15	8.81
profile 19	m2	0.42	2.94	4.50	1.12	8.56
profile MS20	m2	0.45	3.15	4.82	1.20	9.17
profile 32	m2	0.45	3.15	5.02	1.23	9.40
profile 35	m2	0.47	3.29	5.34	1.29	9.92
profile 38A	m2	0.48	3.36	5.27	1.29	9.92
profile 40	m2	0.50	3.50	4.82	1.25	9.57
profile 46	m2	0.50	3.50	5.34	1.33	10.17
profile 60	m2	0.56	3.92	6.02	1.49	11.43
profile 100	m2	0.60	4.20	6.88	1.66	12.74

Wall cladding, fixed to steelwork with self-tapping screws, 0.90mm thick

profile 13.5/3	m2	0.40	2.80	6.12	1.34	10.26
profile 19	m2	0.42	2.94	5.68	1.29	9.91

RATES FOR MEASURED WORK

Aluminium sheeting (cont'd)	Unit	Labour hours	Net labour (£)	Net material (£)	O'heads /profit (£)	Total (£)
profile MS20	m2	0.45	3.15	6.07	1.38	10.60
profile 32	m2	0.45	3.15	6.32	1.42	10.89
profile 35	m2	0.47	3.29	6.74	1.50	11.53
profile 38A	m2	0.48	3.36	6.63	1.50	11.49
profile 40	m2	0.50	3.50	6.07	1.44	11.01
profile 46	m2	0.50	3.50	6.74	1.54	11.78
profile 60	m2	0.56	3.92	7.57	1.72	13.21
profile 100	m2	0.60	4.20	8.65	1.93	14.78
Wall cladding, fixed to steelwork with self-tapping screws, 1.20mm thick						
profile 13.5/3	m2	0.40	2.80	7.93	1.61	12.34
profile 19	m2	0.42	2.94	7.36	1.54	11.84
profile MS20	m2	0.45	3.15	7.86	1.65	12.66
profile 32	m2	0.45	3.15	8.18	1.70	13.03
profile 35	m2	0.47	3.29	8.74	1.80	13.83
profile 38A	m2	0.48	3.36	8.59	1.79	13.74
profile 40	m2	0.50	3.50	7.86	1.70	13.06
profile 46	m2	0.50	3.50	8.74	1.84	14.08
profile 60	m2	0.56	3.92	9.83	2.06	15.81
profile 100	m2	0.60	4.20	11.22	2.31	17.73
Roof cladding, fixed to steelwork with self-tapping screws, 0.70mm thick						
profile 13.5/3	m2	0.35	2.45	4.86	1.10	8.41
profile 19	m2	0.37	2.59	4.50	1.06	8.15
profile MS20	m2	0.39	2.73	4.82	1.13	8.68
profile 32	m2	0.39	2.73	5.02	1.16	8.91
profile 35	m2	0.40	2.80	5.34	1.22	9.36
profile 38A	m2	0.42	2.94	5.27	1.23	9.44
profile 40	m2	0.44	3.08	4.82	1.19	9.09
profile 46	m2	0.44	3.08	5.34	1.26	9.68
profile 60	m2	0.56	3.92	6.02	1.49	11.43
profile 100	m2	0.60	4.20	6.88	1.66	12.74

METAL PROFILED SHEETING

	Unit	Labour hours	Net labour (£)	Net material (£)	O'heads /profit (£)	Total (£)
Roof cladding, fixed to steelwork with self-tapping screws, 0.90mm thick						
profile 13.5/3	m2	0.35	2.45	6.12	1.29	9.86
profile 19	m2	0.37	2.59	5.68	1.24	9.51
profile MS20	m2	0.39	2.73	6.07	1.32	10.12
profile 32	m2	0.39	2.73	6.32	1.36	10.41
profile 35	m2	0.40	2.80	6.74	1.43	10.97
profile 38A	m2	0.42	2.94	6.63	1.44	11.01
profile 40	m2	0.44	3.08	6.07	1.37	10.52
profile 46	m2	0.44	3.08	6.74	1.47	11.29
profile 60	m2	0.56	3.92	7.57	1.72	13.21
profile 100	m2	0.60	4.20	8.65	1.93	14.78
Roof cladding, fixed to steelwork with self-tapping screws, 1.20mm thick						
profile 13.5/3	m2	0.35	2.45	7.93	1.56	11.94
profile 19	m2	0.37	2.59	7.36	1.49	11.44
profile MS20	m2	0.39	2.73	7.86	1.59	12.18
profile 32	m2	0.39	2.73	8.18	1.64	12.55
profile 35	m2	0.40	2.80	8.74	1.73	13.27
profile 38A	m2	0.42	2.94	8.59	1.73	13.26
profile 40	m2	0.44	3.08	7.86	1.64	12.58
profile 46	m2	0.44	3.08	8.74	1.77	13.59
profile 60	m2	0.56	3.92	9.83	2.06	15.81
profile 100	m2	0.60	4.20	11.22	2.31	17.73
Composite floor decking, fixed to steelwork with self-tapping screws, 0.90mm thick						
profile CF46	m2	0.41	2.87	7.09	1.49	11.45
profile CF51	m2	0.43	3.01	8.83	1.78	13.62
profile CF70	m2	0.44	3.08	6.58	1.45	11.11

RATES FOR MEASURED WORK

Steel decking (cont'd)	Unit	Labour hours	Net labour (£)	Net material (£)	O'heads /profit (£)	Total (£)
Composite floor decking, fixed to steelwork with self-tapping screws, 1.20mm thick						
profile CF46	m2	0.41	2.87	9.20	1.81	13.88
profile CF41	m2	0.43	3.01	11.81	2.22	17.04
profile CF70	m2	0.44	3.08	8.52	1.74	13.34
Colour coated galvanized steel roll-formed profiled sheets						
Wall cladding, fixed to steelwork with self-tapping screws, 0.70mm thick, external face HP200, internal standard backing coat						
profile 13.5/3	m2	0.40	2.80	6.76	1.43	10.99
profile 19	m2	0.42	2.94	6.27	1.38	10.59
profile MS20	m2	0.45	3.15	6.69	1.48	11.32
profile 32	m2	0.47	3.29	6.97	1.54	11.80
profile 35	m2	0.47	3.29	7.44	1.61	12.34
profile 38A	m2	0.48	3.36	7.32	1.60	12.28
profile 40	m2	0.50	3.50	6.69	1.53	11.72
profile 46	m2	0.50	3.50	8.20	1.75	13.45
profile 60	m2	0.56	3.92	9.23	1.97	15.12
profile 100	m2	0.60	4.20	10.55	2.21	16.96
Wall cladding, fixed to steelwork with self-tapping screws, 0.55mm thick, external face HP200, internal standard backing coat						
profile 13.5/3	m2	0.40	2.80	6.14	1.34	10.28
profile 19	m2	0.42	2.94	5.70	1.30	9.94
profile MS20	m2	0.45	3.15	6.09	1.39	10.63
profile 32	m2	0.47	3.29	6.34	1.44	11.07
profile 35	m2	0.47	3.29	6.76	1.51	11.56
profile 38A	m2	0.48	3.36	6.66	1.50	11.52
profile 40	m2	0.50	3.50	6.09	1.44	11.03
profile 46	m2	0.50	3.50	7.26	1.61	12.37
profile 60	m2	0.56	3.92	8.17	1.81	13.90

METAL PROFILED SHEETING

	Unit	Labour hours	Net labour (£)	Net material (£)	O'heads /profit (£)	Total (£)

Wall cladding, fixed to steelwork with self-tapping screws, 0.70mm thick, external face Pvf2, internal face standard backing coat

profile 13.5/3	m2	0.42	2.94	7.04	1.50	11.48
profile 19	m2	0.42	2.94	6.54	1.42	10.90
profile MS20	m2	0.45	3.15	6.97	1.52	11.64
profile 32	m2	0.47	3.29	7.27	1.58	12.14
profile 35	m2	0.47	3.29	7.75	1.66	12.70
profile 38A	m2	0.48	3.36	7.63	1.65	12.64
profile 40	m2	0.50	3.50	6.97	1.57	12.04
profile 46	m2	0.50	3.50	8.20	1.75	13.45
profile 60	m2	0.56	3.92	9.23	1.97	15.12
profile 100	m2	0.60	4.20	10.55	2.21	16.96

Wall cladding, fixed to steelwork with self-tapping screws, 0.70mm thick, external face lining enamel, internal face standard backing coat

profile 13.5/3	m2	0.42	2.94	6.42	1.40	10.76
profile 19	m2	0.42	2.94	5.95	1.33	10.22
profile MS20	m2	0.45	3.15	6.36	1.43	10.94
profile 32	m2	0.47	3.29	6.63	1.49	11.41
profile 35	m2	0.47	3.29	8.17	1.72	13.18
profile 38A	m2	0.48	3.36	6.96	1.55	11.87
profile 40	m2	0.50	3.50	6.36	1.48	11.34
profile 46	m2	0.50	3.50	7.39	1.63	12.52
profile 60	m2	0.56	3.92	8.32	1.84	14.08
profile 100	m2	0.60	4.20	9.50	2.06	15.76

Roof cladding, fixed to steelwork with self-tapping screws, 0.70mm thick, external face HP200, internal face standard backing coat

profile 13.5/3	m2	0.33	2.31	6.76	1.36	10.43
profile 19	m2	0.37	2.59	6.27	1.33	10.19
profile MS20	m2	0.39	2.73	6.69	1.41	10.83
profile 32	m2	0.39	2.73	6.97	1.45	11.15

RATES FOR MEASURED WORK

Colour coated steel (cont'd)	Unit	Labour hours	Net labour (£)	Net material (£)	O'heads /profit (£)	Total (£)
profile 35	m2	0.40	2.80	7.44	1.54	11.78
profile 38A	m2	0.42	2.94	7.32	1.54	11.80
profile 40	m2	0.42	2.94	6.69	1.44	11.07
profile 46	m2	0.44	3.08	8.20	1.69	12.97
profile 60	m2	0.52	3.64	9.23	1.93	14.80
profile 100	m2	0.56	3.92	10.55	2.17	16.64

Roof cladding, fixed to steelwork with self-tapping screws, 0.55mm thick, external face HP200, internal face standard backing coat

profile 13.5/3	m2	0.33	2.31	6.14	1.27	9.72
profile 19	m2	0.37	2.59	5.70	1.24	9.53
profile MS20	m2	0.39	2.73	6.09	1.32	10.14
profile 32	m2	0.39	2.73	6.34	1.36	10.43
profile 35	m2	0.40	2.80	6.76	1.43	10.99
profile 38A	m2	0.42	2.94	6.66	1.44	11.04
profile 40	m2	0.42	2.94	6.09	1.35	10.38
profile 46	m2	0.44	3.08	7.26	1.55	11.89
profile 60	m2	0.52	3.64	8.17	1.77	13.58

Roof cladding, fixed to steelwork with self-tapping screws, 0.70mm thick, external face Pvf2, internal face standard backing coat

profile 13.5/3	m2	0.33	2.31	7.04	1.40	10.75
profile 19	m2	0.37	2.59	6.54	1.37	10.50
profile MS20	m2	0.39	2.73	6.97	1.45	11.15
profile 32	m2	0.39	2.73	7.27	1.50	11.50
profile 35	m2	0.40	2.80	7.75	1.58	12.13
profile 38A	m2	0.42	2.94	7.63	1.59	12.16
profile 40	m2	0.42	2.94	6.97	1.49	11.40
profile 46	m2	0.44	3.08	8.20	1.69	12.97
profile 60	m2	0.52	3.64	9.23	1.93	14.80
profile 100	m2	0.56	3.92	10.55	2.17	16.64

METAL PROFILED SHEETING

	Unit	Labour hours	Net labour (£)	Net material (£)	O'heads /profit (£)	Total (£)
Roof cladding, fixed to steelwork with self tapping screws, 0.70mm thick, external face lining enamel, internal face standard backing coat						
profile 13.5/3	m2	0.33	2.31	6.42	1.31	10.04
profile 19	m2	0.37	2.59	5.95	1.28	9.82
profile MS20	m2	0.39	2.73	6.36	1.36	10.45
profile 32	m2	0.39	2.73	6.63	1.40	10.76
profile 35	m2	0.40	2.80	8.17	1.65	12.62
profile 38A	m2	0.42	2.94	6.96	1.48	11.38
profile 40	m2	0.42	2.94	6.36	1.40	10.70
profile 46	m2	0.44	3.08	7.39	1.57	12.04
profile 60	m2	0.52	3.64	7.21	1.63	12.48
profile 100	m2	0.56	3.92	9.50	2.01	15.43
Colour coated galvanized steel pressed profiled sheets						
Wall cladding, fixed to steelwork with self-tapping screws, 0.70mm thick, external face HP200, internal face standard backing coat						
profile PR8	m2	0.42	2.94	8.65	1.74	13.33
profile PM13	m2	0.43	3.01	9.25	1.84	14.10
profile PL19	m2	0.44	3.08	9.56	1.90	14.54
profile PG22	m2	0.45	3.15	10.14	1.99	15.28
profile PS47	m2	0.46	3.22	12.55	2.37	18.14
Wall cladding, fixed to steelwork with self-tapping screws, 0.55mm thick, external face HP200, internal face standard backing coat						
profile PR8	m2	0.42	2.94	7.67	1.59	12.20
profile PM13	m2	0.43	3.01	8.19	1.68	12.88
profile PL19	m2	0.44	3.08	8.46	1.73	13.27
profile PG22	m2	0.45	3.15	8.98	1.82	13.95
profile PS47	m2	0.46	3.22	11.10	2.15	16.47

RATES FOR MEASURED WORK

Colour coated steel (cont'd)	Unit	Labour hours	Net labour (£)	Net material (£)	O'heads /profit (£)	Total (£)
Wall cladding, fixed to steelwork with self-tapping screws, 0.70mm thick, external face Pvf2, internal face standard backing coat						
profile PR8	m2	0.42	2.94	9.62	1.88	14.44
profile PM13	m2	0.43	3.01	10.29	1.99	15.29
profile PL19	m2	0.44	3.08	10.63	2.06	15.77
profile PG22	m2	0.45	3.15	11.27	2.16	16.58
profile PS47	m2	0.46	3.22	13.94	2.57	19.73
Roof cladding, fixed to steelwork with self-tapping screws, 0.70mm thick, external face HP200, internal face standard backing coat						
profile PR8	m2	0.37	2.59	8.65	1.69	12.93
profile PM13	m2	0.39	2.73	9.25	1.80	13.78
profile PL19	m2	0.41	2.87	9.56	1.86	14.29
profile PG22	m2	0.43	3.01	10.14	1.97	15.12
profile PS47	m2	0.44	3.08	12.55	2.34	17.97
Roof cladding, fixed to steelwork with self-tapping screws, 0.55mm thick, external face HP200, internal face standard backing coat						
profile PR8	m2	0.37	2.59	7.67	1.54	11.80
profile PM13	m2	0.39	2.73	8.19	1.64	12.56
profile PL19	m2	0.41	2.87	8.46	1.70	13.03
profile PG22	m2	0.43	3.01	8.98	1.80	13.79
profile PS47	m2	0.44	3.08	11.10	2.13	16.31

METAL PROFILED SHEETING

	Unit	Labour hours	Net labour (£)	Net material (£)	O'heads /profit (£)	Total (£)
Roof cladding, fixed to steelwork with self-tapping screws, 0.70mm thick, external face Pvf2, internal face standard backing coat						
profile PR8	m2	0.37	2.59	9.62	1.83	14.04
profile PM13	m2	0.39	2.73	10.29	1.95	14.97
profile PL19	m2	0.41	2.87	10.63	2.02	15.52
profile PG22	m2	0.43	3.01	11.27	2.14	16.42
profile PS47	m2	0.44	3.08	13.94	2.55	19.57
Colour coated galvanized steel lining systems						
Roll-formed panels, 0.40mm thick, external face lining enamel, internal face standard backing coat						
profile CL3/900	m2	0.28	1.96	3.95	0.89	6.80
profile CL3/914	m2	0.28	1.96	3.90	0.88	6.74
profile CL6/914	m2	0.28	1.96	3.90	0.88	6.74
profile CL3/960	m2	0.28	1.96	3.71	0.85	6.52
profile CL3/1000	m2	0.28	1.96	3.66	0.84	6.46
profile CL3/1016	m2	0.28	1.96	3.60	0.83	6.39
Roll-formed panels, 0.70mm thick, external face lining enamel, internal face standard backing coat						
profile 13.5/3	m2	0.40	2.80	6.42	1.38	10.60
profile 19	m2	0.42	2.94	5.95	1.33	10.22
profile MS20	m2	0.45	3.15	6.36	1.43	10.94
Structural lining tray 0.90mm thick, external face lining enamel, internal face standard backing coat						
profile HL70/400	m2	0.56	3.92	14.18	2.71	20.81

RATES FOR MEASURED WORK

Flashings (cont'd)	Unit	Labour hours	Net labour (£)	Net material (£)	O'heads /profit (£)	Total (£)
Shadow line tray 0.70mm thick, external finish lining enamel, internal face standard backing coat						
up to 408mm total girth	m	0.24	1.68	2.88	0.68	5.24
up to 613mm total girth	m	0.30	2.10	4.33	0.96	7.39
Shadow line tray 0.70mm thick, external face HP200, internal face standard backing coat						
up to 408mm total girth	m	0.24	1.68	3.35	0.75	5.78
up to 613mm total girth	m	0.30	2.10	5.02	1.07	8.19
Insulation						
Polystyrene (EPS)						
profile C19, thickness 62mm	m2	0.15	1.05	4.89	0.89	6.83
profile R62, thickness 55mm	m2	0.15	1.05	7.53	1.29	9.87
profile C32, thickness 53mm	m2	0.15	1.05	7.35	1.26	9.66
profile R38A, thickness 66mm	m2	0.15	1.05	7.32	1.26	9.63
profile C38A, thickness 50mm	m2	0.15	1.05	7.48	1.28	9.81
profile R40, thickness 67mm	m2	0.15	1.05	5.58	0.99	7.62
profile C40, thickness 43mm	m2	0.15	1.05	5.58	0.99	7.62
Rockfibre 100						
profile R32, thickness 67mm	m2	0.18	1.26	15.97	2.58	19.81
profile C32, thickness 54mm	m2	0.18	1.26	15.97	2.58	19.81
profile R38A, thickness 68mm	m2	0.18	1.26	16.37	2.64	20.27
profile C38A, thickness 52mm	m2	0.18	1.26	16.37	2.64	20.27
profile R40, thickness 69mm	m2	0.18	1.26	17.07	2.75	21.08
profile C40, thickness 45mm	m2	0.18	1.26	17.07	2.75	21.08
Flashings						
Aluminium alloy, plain or stucco						
0.70mm thick						
100mm girth	m	0.16	1.12	1.40	0.38	2.90
153mm girth	m	0.17	1.19	1.90	0.46	3.55
204mm girth	m	0.18	1.26	2.52	0.57	4.35

METAL PROFILED SHEETING

	Unit	Labour hours	Net labour (£)	Net material (£)	O'heads /profit (£)	Total (£)
245mm girth	m	0.19	1.33	3.03	0.65	5.01
306mm girth	m	0.20	1.40	3.78	0.78	5.96
350mm girth	m	0.21	1.47	4.90	0.96	7.33
408mm girth	m	0.22	1.54	5.05	0.99	7.58
450mm girth	m	0.23	1.61	6.31	1.19	9.11
500mm girth	m	0.24	1.68	7.00	1.30	9.98
612mm girth	m	0.30	2.10	7.57	1.45	11.12
700mm girth	m	0.35	2.45	9.80	1.84	14.09
800mm girth	m	0.40	2.80	11.21	2.10	16.11
900mm girth	m	0.45	3.15	12.61	2.36	18.12
1000mm girth	m	0.50	3.50	14.02	2.63	20.15
0.90mm thick						
100mm girth	m	0.16	1.12	1.76	0.43	3.31
153mm girth	m	0.17	1.19	2.39	0.54	4.12
204mm girth	m	0.18	1.26	3.18	0.67	5.11
245mm girth	m	0.19	1.33	3.82	0.77	5.92
306mm girth	m	0.20	1.40	4.78	0.93	7.11
350mm girth	m	0.21	1.47	6.18	1.15	8.80
408mm girth	m	0.22	1.54	6.37	1.19	9.10
450mm girth	m	0.23	1.61	7.96	1.44	11.01
500mm girth	m	0.24	1.68	8.84	1.58	12.10
612mm girth	m	0.30	2.10	9.56	1.75	13.41
700mm girth	m	0.35	2.45	12.37	2.22	17.04
800mm girth	m	0.40	2.80	14.14	2.54	19.48
900mm girth	m	0.45	3.15	15.91	2.86	21.92
1000mm girth	m	0.50	3.50	17.67	3.18	24.35
Aluminium alloy, plain						
1.20mm thick						
100mm girth	m	0.16	1.12	2.29	0.51	3.92
153mm girth	m	0.17	1.19	3.11	0.64	4.94
204mm girth	m	0.18	1.26	4.15	0.81	6.22
245mm girth	m	0.19	1.33	4.98	0.95	7.26
306mm girth	m	0.20	1.40	6.21	1.14	8.75
350mm girth	m	0.21	1.47	8.03	1.42	10.92
408mm girth	m	0.22	1.54	8.28	1.47	11.29
450mm girth	m	0.23	1.61	10.34	1.79	13.74
500mm girth	m	0.24	1.68	11.49	1.98	15.15
612mm girth	m	0.30	2.10	12.42	2.18	16.70
700mm girth	m	0.35	2.45	16.09	2.78	21.32
800mm girth	m	0.40	2.80	18.38	3.18	24.36
900mm girth	m	0.45	3.15	20.69	3.58	27.42

RATES FOR MEASURED WORK

Flashings (cont'd)	Unit	Labour hours	Net labour (£)	Net material (£)	O'heads /profit (£)	Total (£)
1000mm girth	m	0.50	3.50	22.97	3.97	30.44
1.60mm thick						
100mm girth	m	0.16	1.12	3.01	0.62	4.75
153mm girth	m	0.17	1.19	4.06	0.79	6.04
204mm girth	m	0.18	1.26	5.43	1.00	7.69
245mm girth	m	0.19	1.33	6.52	1.18	9.03
306mm girth	m	0.20	1.40	8.14	1.43	10.97
350mm girth	m	0.21	1.47	10.54	1.80	13.81
408mm girth	m	0.22	1.54	10.85	1.86	14.25
450mm girth	m	0.23	1.61	13.55	2.27	17.43
500mm girth	m	0.24	1.68	15.05	2.51	19.24
612mm girth	m	0.30	2.10	16.27	2.76	21.13
700mm girth	m	0.35	2.45	21.08	3.53	27.06
800mm girth	m	0.40	2.80	24.09	4.03	30.92
900mm girth	m	0.45	3.15	27.10	4.54	34.79
1000mm girth	m	0.50	3.50	30.11	5.04	38.65
Aluminium alloy, AD80/standard lacquer						
0.70mm thick						
100mm girth	m	0.16	1.12	2.49	0.54	4.15
153mm girth	m	0.17	1.19	3.36	0.68	5.23
204mm girth	m	0.18	1.26	4.48	0.86	6.60
245mm girth	m	0.19	1.33	5.39	1.01	7.73
306mm girth	m	0.20	1.40	6.74	1.22	9.36
350mm girth	m	0.21	1.47	8.73	1.53	11.73
408mm girth	m	0.22	1.54	8.98	1.58	12.10
450mm girth	m	0.23	1.61	11.21	1.92	14.74
500mm girth	m	0.24	1.68	12.46	2.12	16.26
612mm girth	m	0.30	2.10	13.47	2.34	17.91
700mm girth	m	0.35	2.45	17.44	2.98	22.87
800mm girth	m	0.40	2.80	19.93	3.41	26.14
900mm girth	m	0.45	3.15	22.41	3.83	29.39
1000mm girth	m	0.50	3.50	24.92	4.26	32.68
0.90mm thick						
100mm girth	m	0.00	0.00	3.09	0.46	3.55
153mm girth	m	0.00	0.00	4.18	0.63	4.81
204mm girth	m	0.00	0.00	5.56	0.83	6.39
245mm girth	m	0.00	0.00	6.69	1.00	7.69
306mm girth	m	0.00	0.00	8.36	1.25	9.61
350mm girth	m	0.00	0.00	10.83	1.62	12.45

METAL PROFILED SHEETING

	Unit	Labour hours	Net labour (£)	Net material (£)	O'heads /profit (£)	Total (£)
408mm girth	m	0.00	0.00	11.15	1.67	12.82
450mm girth	m	0.00	0.00	13.91	2.09	16.00
500mm girth	m	0.00	0.00	15.46	2.32	17.78
612mm girth	m	0.00	0.00	16.72	2.51	19.23
700mm girth	m	0.00	0.00	21.64	3.25	24.89
800mm girth	m	0.00	0.00	24.74	3.71	28.45
900mm girth	m	0.00	0.00	27.83	4.17	32.00
1000mm girth	m	0.00	0.00	30.92	4.64	35.56

Aluminium alloy, galvanized steel

0.70mm thick
100mm girth	m	0.16	1.12	0.92	0.31	2.35
153mm girth	m	0.17	1.19	1.60	0.42	3.21
204mm girth	m	0.18	1.26	2.13	0.51	3.90
245mm girth	m	0.19	1.33	2.56	0.58	4.47
306mm girth	m	0.20	1.40	3.19	0.69	5.28
350mm girth	m	0.21	1.47	4.29	0.86	6.62
408mm girth	m	0.22	1.54	4.25	0.87	6.66
450mm girth	m	0.23	1.61	5.52	1.07	8.20
500mm girth	m	0.24	1.68	6.14	1.17	8.99
612mm girth	m	0.30	2.10	6.39	1.27	9.76
700mm girth	m	0.35	2.45	8.60	1.66	12.71
800mm girth	m	0.40	2.80	9.83	1.89	14.52
900mm girth	m	0.45	3.15	11.06	2.13	16.34
1000mm girth	m	0.50	3.50	12.29	2.37	18.16

0.90mm thick
100mm girth	m	0.16	1.12	1.16	0.34	2.62
153mm girth	m	0.17	1.19	2.01	0.48	3.68
204mm girth	m	0.18	1.26	2.67	0.59	4.52
245mm girth	m	0.19	1.33	3.21	0.68	5.22
306mm girth	m	0.20	1.40	4.00	0.81	6.21
350mm girth	m	0.21	1.47	5.39	1.03	7.89
408mm girth	m	0.22	1.54	5.34	1.03	7.91
450mm girth	m	0.23	1.61	6.93	1.28	9.82
500mm girth	m	0.24	1.68	7.71	1.41	10.80
612mm girth	m	0.30	2.10	8.01	1.52	11.63
700mm girth	m	0.35	2.45	10.78	1.98	15.21
800mm girth	m	0.40	2.80	12.33	2.27	17.40
900mm girth	m	0.45	3.15	13.87	2.55	19.57
1000mm girth	m	0.50	3.50	15.40	2.83	21.73

RATES FOR MEASURED WORK

Flashings (cont'd)	Unit	Labour hours	Net labour (£)	Net material (£)	O'heads /profit (£)	Total (£)
1.20mm thick						
100mm girth	m	0.16	1.12	1.50	0.39	3.01
153mm girth	m	0.17	1.19	2.59	0.57	4.35
204mm girth	m	0.18	1.26	3.47	0.71	5.44
245mm girth	m	0.19	1.33	4.16	0.82	6.31
306mm girth	m	0.20	1.40	5.20	0.99	7.59
350mm girth	m	0.21	1.47	6.99	1.27	9.73
408mm girth	m	0.22	1.54	6.92	1.27	9.73
450mm girth	m	0.23	1.61	8.99	1.59	12.19
500mm girth	m	0.24	1.68	9.98	1.75	13.41
612mm girth	m	0.30	2.10	10.38	1.87	14.35
700mm girth	m	0.35	2.45	13.97	2.46	18.88
800mm girth	m	0.40	2.80	15.96	2.81	21.57
900mm girth	m	0.45	3.15	17.96	3.17	24.28
1000mm girth	m	0.50	3.50	19.96	3.52	26.98
1.60mm thick						
100mm girth	m	0.16	1.12	1.94	0.46	3.52
130mm girth	m	0.17	1.19	3.36	0.68	5.23
204mm girth	m	0.18	1.26	4.47	0.86	6.59
245mm girth	m	0.19	1.33	5.38	1.01	7.72
306mm girth	m	0.20	1.40	6.73	1.22	9.35
350mm girth	m	0.21	1.47	9.04	1.58	12.09
408mm girth	m	0.22	1.54	8.97	1.58	12.09
450mm girth	m	0.23	1.61	11.63	1.99	15.23
500mm girth	m	0.24	1.68	12.93	2.19	16.80
612mm girth	m	0.30	2.10	13.44	2.33	17.87
700mm girth	m	0.35	2.45	18.08	3.08	23.61
800mm girth	m	0.40	2.80	20.67	3.52	26.99
900mm girth	m	0.45	3.15	23.25	3.96	30.36
1000mm girth	m	0.50	3.50	25.83	4.40	33.73
2.00mm thick						
100mm girth	m	0.16	1.12	2.43	0.53	4.08
130mm girth	m	0.17	1.19	4.94	0.92	7.05
204mm girth	m	0.18	1.26	6.58	1.18	9.02
245mm girth	m	0.19	1.33	7.91	1.39	10.63
306mm girth	m	0.20	1.40	9.88	1.69	12.97
350mm girth	m	0.21	1.47	11.30	1.92	14.69
408mm girth	m	0.22	1.54	13.18	2.21	16.93
450mm girth	m	0.23	1.61	14.53	2.42	18.56
500mm girth	m	0.24	1.68	16.14	2.67	20.49
612mm girth	m	0.30	2.10	19.76	3.28	25.14

METAL PROFILED SHEETING

	Unit	Labour hours	Net labour (£)	Net material (£)	O'heads /profit (£)	Total (£)
700mm girth	m	0.35	2.45	22.61	3.76	28.82
800mm girth	m	0.40	2.80	25.82	4.29	32.91
900mm girth	m	0.45	3.15	29.05	4.83	37.03

Galvanized steel, coloured

0.70mm thick, external face HP200, internal face standard backing coat

75mm girth	m	0.16	1.12	1.84	0.44	3.40
153mm girth	m	0.17	1.19	2.94	0.62	4.75
204mm girth	m	0.18	1.26	3.93	0.78	5.97
245mm girth	m	0.19	1.33	4.71	0.91	6.95
306mm girth	m	0.20	1.40	5.89	1.09	8.38
350mm girth	m	0.21	1.47	8.61	1.51	11.59
408mm girth	m	0.22	1.54	7.85	1.41	10.80
450mm girth	m	0.23	1.61	11.08	1.90	14.59
500mm girth	m	0.24	1.68	12.31	2.10	16.09
612mm girth	m	0.30	2.10	11.77	2.08	15.95
700mm girth	m	0.35	2.45	17.23	2.95	22.63
800mm girth	m	0.40	2.80	19.69	3.37	25.86
900mm girth	m	0.45	3.15	22.14	3.79	29.08
1000mm girth	m	0.50	3.50	24.62	4.22	32.34

0.55mm thick, external face HP200, internal face standard backing coat

75mm girth	m	0.16	1.12	1.66	0.42	3.20
153mm girth	m	0.17	1.19	2.64	0.57	4.40
204mm girth	m	0.18	1.26	3.52	0.72	5.50
245mm girth	m	0.19	1.33	4.22	0.83	6.38
306mm girth	m	0.20	1.40	5.27	1.00	7.67
350mm girth	m	0.21	1.47	7.72	1.38	10.57
408mm girth	m	0.22	1.54	7.04	1.29	9.87
450mm girth	m	0.23	1.61	9.92	1.73	13.26
500mm girth	m	0.24	1.68	11.03	1.91	14.62
612mm girth	m	0.30	2.10	10.54	1.90	14.54
700mm girth	m	0.35	2.45	15.42	2.68	20.55
800mm girth	m	0.40	2.80	17.63	3.06	23.49
900mm girth	m	0.45	3.15	19.83	3.45	26.43
1000mm girth	m	0.50	3.50	22.04	3.83	29.37

RATES FOR MEASURED WORK

Flashings (cont'd)	Unit	Labour hours	Net labour (£)	Net material (£)	O'heads /profit (£)	Total (£)
0.70mm thick, external face Pvf2, internal face standard backing coat						
75mm girth	m	0.16	1.12	1.84	0.44	3.40
153mm girth	m	0.17	1.19	2.94	0.62	4.75
204mm girth	m	0.18	1.26	3.93	0.78	5.97
245mm girth	m	0.19	1.33	4.71	0.91	6.95
306mm girth	m	0.20	1.40	5.89	1.09	8.38
350mm girth	m	0.21	1.47	8.61	1.51	11.59
408mm girth	m	0.22	1.54	7.85	1.41	10.80
450mm girth	m	0.23	1.61	11.08	1.90	14.59
500mm girth	m	0.30	2.10	12.31	2.16	16.57
612mm girth	m	0.35	2.45	11.77	2.13	16.35
700mm girth	m	0.40	2.80	17.23	3.00	23.03
800mm girth	m	0.45	3.15	19.69	3.43	26.27
900mm girth	m	0.50	3.50	22.14	3.85	29.49
1000mm girth	m	0.00	0.00	24.62	3.69	28.31
0.70mm thick, external face lining enamel, internal face standard backing coat						
75mm girth	m	0.16	1.12	1.50	0.39	3.01
153mm girth	m	0.17	1.19	2.39	0.54	4.12
204mm girth	m	0.18	1.26	3.19	0.67	5.12
245mm girth	m	0.19	1.33	3.83	0.77	5.93
306mm girth	m	0.20	1.40	4.80	0.93	7.13
350mm girth	m	0.21	1.47	7.01	1.27	9.75
408mm girth	m	0.22	1.54	6.39	1.19	9.12
450mm girth	m	0.23	1.61	9.02	1.59	12.22
500mm girth	m	0.24	1.68	10.01	1.75	13.44
612mm girth	m	0.30	2.10	9.58	1.75	13.43
700mm girth	m	0.35	2.45	14.03	2.47	18.95
800mm girth	m	0.40	2.80	16.02	2.82	21.64
900mm girth	m	0.45	3.15	18.03	3.18	24.36
1000mm girth	m	0.50	3.50	20.03	3.53	27.06
0.40mm thick, external face lining enamel, internal face standard backing coat						
75mm girth	m	0.16	1.12	1.03	0.32	2.47
153mm girth	m	0.17	1.19	2.09	0.49	3.77
204mm girth	m	0.18	1.26	2.80	0.61	4.67
245mm girth	m	0.19	1.33	3.36	0.70	5.39
306mm girth	m	0.20	1.40	4.20	0.84	6.44

METAL PROFILED SHEETING

	Unit	Labour hours	Net labour (£)	Net material (£)	O'heads /profit (£)	Total (£)
350mm girth	m	0.21	1.47	4.80	0.94	7.21
408mm girth	m	0.22	1.54	5.60	1.07	8.21
450mm girth	m	0.23	1.61	6.17	1.17	8.95
500mm girth	m	0.24	1.68	6.86	1.28	9.82
612mm girth	m	0.30	2.10	8.39	1.57	12.06
700mm girth	m	0.35	2.45	9.61	1.81	13.87
800mm girth	m	0.40	2.80	10.97	2.07	15.84
900mm girth	m	0.45	3.15	12.35	2.32	17.82
1000mm girth	m	0.50	3.50	13.71	2.58	19.79
0.90mm thick, external face lining enamel, internal face standard backing coat						
75mm girth	m	0.16	1.25	2.02	0.48	3.65
153mm girth	m	0.17	1.19	4.13	0.80	6.12
204mm girth	m	0.19	1.33	5.49	1.02	7.84
245mm girth	m	0.20	1.40	6.59	1.20	9.19
306mm girth	m	0.21	1.47	8.23	1.46	11.16
350mm girth	m	0.22	1.54	9.43	1.65	12.62
408mm girth	m	0.23	1.61	10.98	1.89	14.48
450mm girth	m	0.24	1.68	12.03	2.06	15.77
500mm girth	m	0.30	2.10	13.46	2.33	17.89
612mm girth	m	0.35	2.45	16.47	2.84	21.76

RATES FOR MEASURED WORK

	Unit	Labour hours	Net labour (£)	Net material (£)	O'heads /profit (£)	Total (£)
H41 GLASS REINFORCED CLADDING						
Corrolux vinyl sheets to BS4203 translucent sheeting 1.3mm thick, lapped one corrugation at sides and 150mm at ends fixed with screws and washers to timber purlins						
Profile 3 sheets	m2	0.75	5.25	7.62	1.93	14.80
Profile 6 sheets	m2	0.70	4.90	11.30	2.43	18.63
Corrolux vinyl sheets to BS4203 translucent sheeting 1.3mm thick, lapped one corrugation at sides and 150mm at ends fixed with hook bolts and washers to steel purlins						
Profile 3 sheets	m2	0.85	5.95	7.62	2.04	15.61
Profile 6 sheets	m2	0.80	5.60	11.30	2.54	19.44

ROOF TILING

	Unit	Labour hours	Net labour (£)	Net material (£)	O'heads /profit (£)	Total (£)

H60 CLAY/CONCRETE ROOF TILING

Marley Plain granuled or smooth finish tiles size 267 x 165mm, 65mm lap, 35 degrees pitch, type 1F reinforced underlay

Battens size 38 x 19mm

gauge 100mm	m2	1.73	12.11	20.13	4.84	37.08
gauge 95mm	m2	1.75	12.25	21.37	5.04	38.66
gauge 90mm	m2	1.78	12.46	22.67	5.27	40.40

Battens size 38 x 25mm

gauge 100mm	m2	1.83	12.81	20.34	4.97	38.12
gauge 95mm	m2	1.85	12.95	21.59	5.18	39.72
gauge 90mm	m2	1.88	13.16	22.91	5.41	41.48

Extra for

nailing every tile with aluminium nails	m2	0.30	2.10	0.21	0.35	2.66
interlocking dry verge system	m	0.20	1.40	7.43	1.32	10.15
verge, 150mm wide plain tile undercloak	m	0.20	1.40	1.02	0.36	2.78
double course at eaves	m	0.35	2.45	1.66	0.62	4.73
segmental ridge tile	m	0.42	2.94	6.33	1.39	10.66
segmental monoridge tiles	m	0.62	4.34	10.06	2.16	16.56
valley trough tiles	m	0.62	4.34	6.29	1.59	12.22
segmental hip tiles	m	0.62	4.34	6.29	1.59	12.22
bonnet hip tiles	m	0.80	5.60	6.29	1.78	13.67
Marley eaves vent system	m	0.40	2.80	10.27	1.96	15.03
ventilated ridge terminal	nr	0.60	4.20	29.77	5.10	39.07
gas vent terminal	nr	0.60	4.20	46.31	7.58	58.09
soil vent terminal	nr	0.60	4.20	28.67	4.93	37.80
cutting	m	0.20	1.40	0.00	0.21	1.61
holes for pipes	nr	0.35	2.45	0.00	0.37	2.82

RATES FOR MEASURED WORK

	Unit	Labour hours	Net labour (£)	Net material (£)	O'heads /profit (£)	Total (£)
Marley Plain Premium granuled or smooth finish tiles size 267 x 165mm, 65mm lap, 35 degrees pitch, type 1F type reinforced underlay						
Battens size 38 x 19mm						
gauge 100mm	m2	1.73	12.11	22.65	5.21	39.97
gauge 95mm	m2	1.75	12.25	23.93	5.43	41.61
gauge 90mm	m2	1.78	12.46	25.22	5.65	43.33
Battens size 38 x 25mm						
gauge 100mm	m2	1.83	12.81	22.86	5.35	41.02
gauge 95mm	m2	1.85	12.95	24.14	5.56	42.65
gauge 90mm	m2	1.88	13.16	25.43	5.79	44.38
Extra for						
nailing every tile with aluminium nails	m2	0.30	2.10	1.26	0.50	3.86
interlocking dry verge system	m	0.20	1.40	7.43	1.32	10.15
verge, 150mm plain tile undercloak	m	0.20	1.40	1.02	0.36	2.78
double course at eaves	m	0.35	2.45	1.66	0.62	4.73
segmental ridge tile	m	0.42	2.94	6.33	1.39	10.66
segmental monoridge tiles	m	0.62	4.34	10.06	2.16	16.56
valley trough tiles	m	0.62	4.34	6.29	1.59	12.22
segmental hip tiles	m	0.62	4.34	6.29	1.59	12.22
bonnet hip tiles	m	0.80	5.60	6.29	1.78	13.67
Marley eaves vent system	m	0.40	2.80	10.27	1.96	15.03
ventilated ridge terminal	nr	0.60	4.20	29.77	5.10	39.07
gas vent terminal	nr	0.60	4.20	46.31	7.58	58.09
soil vent terminal	nr	0.60	4.20	28.67	4.93	37.80
cutting	m	0.20	1.40	0.00	0.21	1.61
holes for pipes	nr	0.35	2.45	0.00	0.37	2.82

ROOF TILING

	Unit	Labour hours	Net labour (£)	Net material (£)	O'heads /profit (£)	Total (£)
Marley Feature granuled or smooth finish tiles size 267 x 165mm, 35mm lap, vertical, type 1F reinforced underlay						
Battens size 38 x 19mm						
gauge 115mm	m2	1.73	12.11	24.81	5.54	42.46
gauge 110mm	m2	1.66	11.62	26.14	5.66	43.42
Battens size 38 x 25mm						
gauge 115mm	m2	1.83	12.81	24.99	5.67	43.47
gauge 110mm	m2	1.75	12.25	24.87	5.57	42.69
Extra for						
nailing every tile with aluminium nails	m2	0.28	1.96	1.16	0.47	3.59
interlocking dry verge system	m	0.20	1.40	7.43	1.32	10.15
verge plain tile undercloak	m	0.20	1.40	1.02	0.36	2.78
double course at eaves	m	0.35	2.45	2.42	0.73	5.60
ventilated ridge terminal	nr	0.60	4.20	29.77	5.10	39.07
gas vent terminal	nr	0.60	4.20	46.31	7.58	58.09
soil vent terminal	nr	0.60	4.20	28.67	4.93	37.80
cutting	m	0.20	1.40	0.00	0.21	1.61
holes for pipes	nr	0.35	2.45	0.00	0.37	2.82
Marley Ludlow Plus smooth finish tiles size 387 x 229mm, battens size 38 x 25mm, type 1F reinforced underlay						
75mm lap, pitch 25 to 44 degrees	m2	0.93	6.51	9.01	2.33	17.85
100mm lap, pitch 22 to 44 degrees	m2	1.02	7.14	9.08	2.43	18.65
Extra for						
nailing every tile with aluminium nails	m2	0.09	0.63	0.36	0.15	1.14
interlocking dry verge system	m	0.20	1.40	5.12	0.98	7.50
verge, 150mm wide plain tile undercloak	m	0.20	1.40	1.02	0.36	2.78

RATES FOR MEASURED WORK

Marley roof tiles (cont'd)	Unit	Labour hours	Net labour (£)	Net material (£)	O'heads /profit (£)	Total (£)
segmental ridge tiles	m	0.42	2.94	6.33	1.39	10.66
segmental monoridge tiles	m	0.62	4.34	10.06	2.16	16.56
dry ridge system	m	0.50	3.50	6.24	1.46	11.20
valley trough tiles	m	0.62	4.34	6.29	1.59	12.22
segmental hip tiles	m	0.62	4.34	6.29	1.59	12.22
Marley eaves vent system	m	0.40	2.80	10.27	1.96	15.03
ventilated ridge terminal	nr	0.60	4.20	29.77	5.10	39.07
gas vent terminal	nr	0.60	4.20	46.31	7.58	58.09
soil vent terminal	nr	0.60	4.20	28.67	4.93	37.80
cutting	nr	0.20	1.40	0.00	0.21	1.61
holes for pipes	nr	0.35	2.45	0.00	0.37	2.82

Marley Ludlow Plus granuled finish tiles size 387 x 229mm, battens size 38 x 25mm, type 1F reinforced underlay, 75mm lap

pitch 30 to 44 degrees	m2	0.89	6.23	8.16	2.16	16.55

Extra for

nailing every tile with aluminium nails	m2	0.09	0.63	0.36	0.15	1.14
interlocking dry verge system	m	0.20	1.40	5.12	0.98	7.50
verge, 150mm wide plain tile undercloak	m	0.20	1.40	1.02	0.36	2.78
segmental ridge tiles	m	0.42	2.94	6.33	1.39	10.66
segmental monoridge tiles	m	0.62	4.34	10.06	2.16	16.56
dry ridge system	m	0.50	3.50	6.24	1.46	11.20
valley trough tiles	m	0.62	4.34	6.29	1.59	12.22
segmental hip tiles	m	0.62	4.34	6.29	1.59	12.22
Marley eaves vent system	m	0.40	2.80	10.27	1.96	15.03
ventilated ridge terminal	nr	0.60	4.20	29.77	5.10	39.07
gas vent terminal	nr	0.60	4.20	46.31	7.58	58.09
soil vent terminal	nr	0.60	4.20	28.67	4.93	37.80
cutting	nr	0.20	1.40	0.00	0.21	1.61
holes for pipes	nr	0.35	2.45	0.00	0.37	2.82

ROOF TILING

	Unit	Labour hours	Net labour (£)	Net material (£)	O'heads /profit (£)	Total (£)
Marley Anglia Plus smooth finish tiles size 387 x 230mm, battens size 38 x 25mm, type 1F reinforced underlay						
75mm lap, pitch 30 to 44 degrees	m2	1.03	7.21	9.65	2.53	19.39
100mm lap, pitch 25 to 44 degrees	m2	1.00	7.00	10.42	2.61	20.03
Extra for						
nailing every tile with aluminium nails	m2	0.08	0.56	0.34	0.14	1.04
interlocking dry verge system	m	0.20	1.40	5.12	0.98	7.50
verge, 150mm wide plain tile undercloak	m	0.20	1.40	1.02	0.36	2.78
segmental ridge tiles	m	0.42	2.94	6.33	1.39	10.66
segmental monoridge tiles	m	0.62	4.34	10.06	2.16	16.56
dry ridge system	m	0.50	3.50	6.24	1.46	11.20
valley trough tiles	m	0.62	4.34	6.29	1.59	12.22
segmental hip tiles	m	0.62	4.34	6.29	1.59	12.22
Marley eaves vent system	m	0.40	2.80	10.27	1.96	15.03
ventilated ridge terminal	nr	0.60	4.20	29.77	5.10	39.07
gas vent terminal	nr	0.60	4.20	46.31	7.58	58.09
soil vent terminal	nr	0.60	4.20	28.67	4.93	37.80
cutting	m	0.20	1.40	0.00	0.21	1.61
holes for pipes	nr	0.35	2.45	0.00	0.37	2.82
Marley Anglia Plus granule finish tiles size 387 x 230mm, battens size 38 x 25mm, type 1F reinforced underlay						
75mm lap, pitch 30 to 44 degrees	m2	1.03	7.21	9.65	2.53	19.39
Extra for						
nailing every tile with aluminium nails	m2	0.08	0.56	0.34	0.14	1.04
interlocking dry verge system	m	0.20	1.40	5.12	0.98	7.50
verge, 150mm wide plain tile undercloak	m	0.20	1.40	1.02	0.36	2.78
segmental monoridge tiles	m	0.62	4.34	10.06	2.16	16.56
dry ridge system	m	0.50	3.50	6.24	1.46	11.20

RATES FOR MEASURED WORK

Marley roof tiles (cont'd)	Unit	Labour hours	Net labour (£)	Net material (£)	O'heads /profit (£)	Total (£)
valley trough tiles	m	0.62	4.34	6.29	1.59	12.22
segmental hip tiles	m	0.62	4.34	6.29	1.59	12.22
Marley eaves vent system	m	0.40	2.80	10.27	1.96	15.03
ventilated ridge terminal	nr	0.60	4.20	29.77	5.10	39.07
gas vent terminal	nr	0.60	4.20	46.31	7.58	58.09
soil vent terminal	nr	0.60	4.20	28.67	4.93	37.80
cutting	m	0.20	1.40	0.00	0.21	1.61
holes for pipes	nr	0.35	2.45	0.00	0.37	2.82

Marley Anglia Plus Premium smooth finish tiles size 387 x 230mm, battens size 38 x 25mm, type 1F reinforced underlay

75mm lap, pitch 30 to 44 degrees	m2	0.92	6.44	9.65	2.41	18.50
100mm lap, pitch 25 to 44 degrees	m2	1.00	7.00	10.34	2.60	19.94

Extra for

nailing every tile with aluminium nails	m2	0.08	0.56	0.34	0.14	1.04
interlocking dry verge system	m	0.20	1.40	5.12	0.98	7.50
verge, 150mm wide plain tile undercloak	m	0.20	1.40	1.02	0.36	2.78
segmental ridge tiles	m	0.42	2.94	6.33	1.39	10.66
segmental monoridge tiles	m	0.62	4.34	10.06	2.16	16.56
dry ridge system	m	0.50	3.50	6.24	1.46	11.20
valley trough tiles	m	0.62	4.34	6.29	1.59	12.22
segmental hip tiles	m	0.62	4.34	6.29	1.59	12.22
Marley eaves vent system	m	0.40	2.80	10.27	1.96	15.03
ventilated ridge terminal	nr	0.60	4.20	29.77	5.10	39.07
gas vent terminal	nr	0.60	4.20	46.31	7.58	58.09
soil vent terminal	nr	0.60	4.20	28.67	4.93	37.80
cutting	m	0.20	1.40	0.00	0.21	1.61
holes for pipes	nr	0.35	2.45	0.00	0.37	2.82

ROOF TILING

	Unit	Labour hours	Net labour (£)	Net material (£)	O'heads /profit (£)	Total (£)
Marley Anglia Plus Premium granule finish tiles size 387 x 230mm, battens size 38 x 25mm, type 1F reinforced underlay						
75mm lap, pitch 30 to 44 degrees	m2	0.02	0.14	9.65	1.47	11.26
Extra for						
nailing every tile with aluminium nails	m2	0.08	0.56	0.34	0.14	1.04
interlocking dry verge system	m	0.20	1.40	5.12	0.98	7.50
verge, 150mm wide plain tile undercloak	m	0.20	1.40	1.02	0.36	2.78
segmental ridge tiles	m	0.42	2.94	6.33	1.39	10.66
segmental monoridge tiles	m	0.62	4.34	10.06	2.16	16.56
dry ridge system	m	0.50	3.50	10.44	2.09	16.03
valley trough tiles	m	0.62	4.34	6.29	1.59	12.22
segmental hip tiles	m	0.62	4.34	6.29	1.59	12.22
Marley eaves vent system	m	0.40	2.80	10.27	1.96	15.03
ventilated ridge terminal	nr	0.60	4.20	29.77	5.10	39.07
gas vent terminal	nr	0.60	4.20	46.31	7.58	58.09
soil vent terminal	nr	0.60	4.20	28.67	4.93	37.80
cutting	m	0.20	1.40	0.00	0.21	1.61
holes for pipes	nr	0.35	2.45	0.00	0.37	2.82
Marley Ludlow Major smooth finish tiles size 420 x 330mm, battens size 38 x 25mm, type 1F reinforced underlay						
75mm lap, pitch 25 to 44 degrees	m2	0.92	6.44	8.41	2.23	17.08
100mm lap, pitch 22.5 to 44 degrees	m2	0.92	6.44	8.99	2.31	17.74
Extra for						
nailing every tile with aluminium nails	m2	0.05	0.35	0.21	0.08	0.64
interlocking dry verge system	m	0.20	1.40	5.12	0.98	7.50
verge, 150mm wide plain tile undercloak	m	0.20	1.40	1.02	0.36	2.78
segmental ridge tiles	m	0.42	2.94	6.33	1.39	10.66

RATES FOR MEASURED WORK

Marley roof tiles (cont'd)	Unit	Labour hours	Net labour (£)	Net material (£)	O'heads /profit (£)	Total (£)
segmental monoridge tiles	m	0.62	4.34	10.06	2.16	16.56
dry ridge system	m	0.50	3.50	6.24	1.46	11.20
valley trough tiles	m	0.62	4.34	6.29	1.59	12.22
segmental hip tiles	m	0.62	4.34	6.29	1.59	12.22
Marley eaves vent system	m	0.40	2.80	10.27	1.96	15.03
ventilated ridge terminal	nr	0.60	4.20	29.77	5.10	39.07
gas vent terminal	nr	0.60	4.20	46.31	7.58	58.09
soil vent terminal	nr	0.60	4.20	28.67	4.93	37.80
cutting	m	0.20	1.40	0.00	0.21	1.61
holes for pipes	nr	0.35	2.45	0.00	0.37	2.82

Marley Ludlow Major granule finish tiles size 420 x 330mm, battens size 38 x 25mm, type 1F reinforced underlay

	Unit	Labour hours	Net labour (£)	Net material (£)	O'heads /profit (£)	Total (£)
75mm lap, pitch 30 to 44 degrees	m2	0.82	5.74	8.41	2.12	16.27

Extra for

	Unit	Labour hours	Net labour (£)	Net material (£)	O'heads /profit (£)	Total (£)
nailing every tile with aluminium nails	m2	0.05	0.35	0.21	0.08	0.64
interlocking dry verge system	m	0.20	1.40	5.12	0.98	7.50
verge, 150mm wide plain tile undercloak	m	0.20	1.40	1.02	0.36	2.78
dry verge system with white PVC interlocking units	m	0.20	1.40	7.43	1.32	10.15
segmental ridge tiles	m	0.42	2.94	6.33	1.39	10.66
segmental monoridge tiles	m	0.62	4.34	10.06	2.16	16.56
dry ridge system	m	0.50	3.50	6.24	1.46	11.20
valley trough tiles	m	0.62	4.34	6.29	1.59	12.22
segmental hip tiles	m	0.62	4.34	6.29	1.59	12.22
Marley eaves vent system	m	0.40	2.80	10.27	1.96	15.03
ventilated ridge terminal	nr	0.60	4.20	29.77	5.10	39.07
gas vent terminal	nr	0.60	4.20	46.31	7.58	58.09
soil vent terminal	nr	0.60	4.20	28.67	4.93	37.80
cutting	m	0.20	1.40	0.00	0.21	1.61
holes for pipes	nr	0.35	2.45	0.00	0.37	2.82

ROOF TILING

	Unit	Labour hours	Net labour (£)	Net material (£)	O'heads /profit (£)	Total (£)
Marley Mendip smooth finish tiles size 420 x 330mm, battens size 38 x 25mm, type 1F reinforced underlay						
75mm lap, pitch 22.5 to 44 degrees	m2	0.82	5.74	9.17	2.24	17.15
Extra for						
nailing every tile with aluminium nails	m2	0.05	0.35	0.21	0.08	0.64
interlocking dry verge system	m	0.20	1.40	4.73	0.92	7.05
verge, 150mm wide plain tile undercloak	m	0.20	1.40	1.02	0.36	2.78
segmental ridge tiles	m	0.42	2.94	6.33	1.39	10.66
segmental monoridge tiles	m	0.62	4.34	10.06	2.16	16.56
dry ridge system	m	0.50	3.50	6.24	1.46	11.20
valley trough tiles	m	0.62	4.34	6.29	1.59	12.22
segmental hip tiles	m	0.62	4.34	6.29	1.59	12.22
Marley eaves vent system	m	0.40	2.80	10.27	1.96	15.03
ventilated ridge terminal	nr	0.60	4.20	29.77	5.10	39.07
gas vent terminal	nr	0.60	4.20	46.31	7.58	58.09
soil vent terminal	nr	0.60	4.20	28.67	4.93	37.80
cutting	m	0.20	1.40	0.00	0.21	1.61
holes for pipes	nr	0.35	2.45	0.00	0.37	2.82
Marley Mendip granule finish tiles size 420 x 330mm, battens size 38 x 25mm, type 1F reinforced underlay						
75mm lap, pitch 30 to 44 degrees	m2	0.83	5.81	9.17	2.25	17.23
100mm lap, pitch 25 to 44 degrees	m2	0.91	6.37	9.83	2.43	18.63
Extra for						
nailing every tile with aluminium nails	m2	0.05	0.35	0.21	0.08	0.64
interlocking dry verge system	m	0.20	1.40	4.73	0.92	7.05
dry verge system with white PVC interlocking units	m	0.20	1.40	6.33	1.16	8.89
segmental ridge tiles	m	0.42	2.94	10.06	1.95	14.95
segmental monoridge tiles	m	0.62	4.34	6.24	1.59	12.17

RATES FOR MEASURED WORK

Marley roof tiles (cont'd)	Unit	Labour hours	Net labour (£)	Net material (£)	O'heads /profit (£)	Total (£)
dry ridge system	m	0.50	3.50	6.24	1.46	11.20
valley trough tiles	m	0.62	4.34	6.29	1.59	12.22
segmental hip tiles	m	0.62	4.34	6.29	1.59	12.22
Marley eaves vent system	m	0.40	2.80	10.27	1.96	15.03
ventilated ridge terminal	m	0.60	4.20	29.77	5.10	39.07
gas vent terminal	nr	0.60	4.20	46.31	7.58	58.09
soil vent terminal	nr	0.60	4.20	28.67	4.93	37.80
cutting	m	0.20	1.40	0.00	0.21	1.61
holes for pipes	nr	0.35	2.45	0.00	0.37	2.82

Marley Mendip Premium granule finish tiles size 420 x 330mm, battens size 38 x 25mm, type 1F reinforced underlay

75mm lap, pitch 30 to 44 degrees	m2	0.83	5.81	11.01	2.52	19.34
100mm lap, pitch 25 to 44 degrees	m2	0.91	6.37	10.26	2.49	19.12

Extra for

nailing every tile with aluminium nails	m2	0.05	0.35	0.21	0.08	0.64
interlocking dry verge system	m	0.20	1.40	4.73	0.92	7.05
verge, 150mm wide plain tile undercloak	m	0.20	1.40	1.02	0.36	2.78
segmental ridge tiles	m	0.42	2.94	6.33	1.39	10.66
segmental monoridge tiles	m	0.62	4.34	10.06	2.16	16.56
dry ridge system	m	0.50	3.50	6.24	1.46	11.20
valley trough tiles	m	0.62	4.34	6.29	1.59	12.22
segmental hip tiles	m	0.62	4.34	6.29	1.59	12.22
Marley eaves vent system	m	0.40	2.80	10.27	1.96	15.03
ventilated ridge terminal	nr	0.60	4.20	29.77	5.10	39.07
gas vent terminal	nr	0.60	4.20	46.31	7.58	58.09
soil vent terminal	nr	0.60	4.20	28.67	4.93	37.80
cutting	m	0.20	1.40	0.00	0.21	1.61
holes for pipes	nr	0.35	2.45	0.00	0.37	2.82

ROOF TILING

	Unit	Labour hours	Net labour (£)	Net material (£)	O'heads /profit (£)	Total (£)
Marley Mendip Premium granule finish tiles size 420 x 330mm, battens size 38 x 25mm, type 1F reinforced underlay						
75mm lap, pitch 30 to 44 degrees	m2	0.83	5.81	11.01	2.52	19.34
100mm lap, pitch 25 to 44 degrees	m2	0.91	6.37	10.26	2.49	19.12
Extra for						
nailing every tile with aluminium nails	m2	0.05	0.35	0.21	0.08	0.64
interlocking dry verge system	m	0.20	1.40	4.73	0.92	7.05
verge, 150mm wide plain tile undercloak	m	0.20	1.40	1.02	0.36	2.78
segmental ridge tiles	m	0.42	2.94	6.33	1.39	10.66
segmental monoridge tiles	m	0.62	4.34	10.06	2.16	16.56
dry ridge system	m	0.50	3.50	6.24	1.46	11.20
valley trough tiles	m	0.62	4.34	6.29	1.59	12.22
segmental hip tiles	m	0.62	4.34	6.29	1.59	12.22
Marley eaves vent system	m	0.40	2.80	10.27	1.96	15.03
ventilated ridge terminal	nr	0.60	4.20	29.77	5.10	39.07
gas vent terminal	nr	0.60	4.20	46.31	7.58	58.09
soil vent terminal	nr	0.60	4.20	28.67	4.93	37.80
cutting	m	0.20	1.40	0.00	0.21	1.61
holes for pipes	nr	0.35	2.45	0.00	0.37	2.82
Marley Double Roman smooth finish tiles size 420 x 330mm, battens size 38 x 25mm, type 1F reinforced underlay						
75mm lap, pitch 25 to 44 degrees	m2	0.83	5.81	8.17	2.10	16.08
100mm lap, pitch 22.5 to 44 degrees	m2	0.91	6.37	8.73	2.27	17.37
Extra for						
nailing every tile with aluminium nails	m2	0.05	0.35	0.21	0.08	0.64
interlocking dry verge system	m	0.20	1.40	4.73	0.92	7.05

RATES FOR MEASURED WORK

Marley roof tiles (cont'd)	Unit	Labour hours	Net labour (£)	Net material (£)	O'heads /profit (£)	Total (£)
verge, 150mm wide plain tile undercloak	m	0.20	1.40	1.02	0.36	2.78
segmental ridge tiles	m	0.42	2.94	6.33	1.39	10.66
segmental monoridge tiles	m	0.62	4.34	10.06	2.16	16.56
dry ridge system	m	0.50	3.50	6.24	1.46	11.20
valley trough tiles	m	0.62	4.34	6.29	1.59	12.22
segmental hip tiles	m	0.62	4.34	6.29	1.59	12.22
Marley eaves vent system	m	0.40	2.80	10.27	1.96	15.03
ventilated ridge terminal	nr	0.60	4.20	29.77	5.10	39.07
gas vent terminal	nr	0.60	4.20	46.31	7.58	58.09
soil vent terminal	nr	0.60	4.20	28.67	4.93	37.80
cutting	m	0.20	1.40	0.00	0.21	1.61
holes for pipes	nr	0.35	2.45	0.00	0.37	2.82

Marley Double Roman granule finish tiles size 420 x 330mm, battens size 38 x 25mm, type 1F reinforced underlay

75mm lap, pitch 30 to 44 degrees	m2	0.83	5.81	8.17	2.10	16.08

Extra for

nailing every tile with aluminium nails	m2	0.05	0.35	0.21	0.08	0.64
interlocking dry verge system	m	0.20	1.40	4.73	0.92	7.05
verge, 150mm wide plain tile undercloak	m	0.20	1.40	1.02	0.36	2.78
segmental ridge tiles	m	0.42	2.94	6.33	1.39	10.66
segmental monoridge tiles	m	0.62	4.34	10.06	2.16	16.56
dry ridge system	m	0.50	3.50	6.24	1.46	11.20
valley trough tiles	m	0.62	4.34	6.29	1.59	12.22
segmental hip tiles	m	0.62	4.34	6.29	1.59	12.22
Marley eaves vent system	m	0.40	2.80	10.27	1.96	15.03
ventilated ridge terminal	nr	0.60	4.20	29.77	5.10	39.07
gas vent terminal	nr	0.60	4.20	46.31	7.58	58.09
soil vent terminal	nr	0.60	4.20	28.67	4.93	37.80
cutting	m	0.20	1.40	0.00	0.21	1.61
holes for pipes	nr	0.35	2.45	0.00	0.37	2.82

ROOF TILING

	Unit	Labour hours	Net labour (£)	Net material (£)	O'heads /profit (£)	Total (£)

Marley Double Roman Premium smooth finish tiles size 420 x 330mm, battens size 38 x 25mm, type 1F reinforced underlay

75mm lap, pitch 25 to 44 degrees	m2	0.82	5.74	9.23	2.25	17.22
100mm lap, pitch 22.5 to 44 degrees	m2	0.83	5.81	9.87	2.35	18.03

Extra for

nailing every tile with aluminium nails	m2	0.05	0.35	0.21	0.08	0.64
interlocking dry verge system	m	0.20	1.40	4.73	0.92	7.05
verge, 150mm wide plain tile undercloak	m	0.20	1.40	1.02	0.36	2.78
segmental ridge tiles	m	0.42	2.94	6.33	1.39	10.66
segmental monoridge tiles	m	0.62	4.34	10.06	2.16	16.56
dry ridge system	m	0.50	3.50	6.24	1.46	11.20
valley trough tiles	m	0.62	4.34	6.29	1.59	12.22
segmental hip tiles	m	0.62	4.34	6.29	1.59	12.22
Marley eaves vent system	m	0.40	2.80	10.27	1.96	15.03
ventilated ridge terminal	nr	0.60	4.20	29.77	5.10	39.07
gas vent terminal	nr	0.60	4.20	46.31	7.58	58.09
soil vent terminal	nr	0.60	4.20	28.67	4.93	37.80
cutting	m	0.20	1.40	0.00	0.21	1.61
holes for pipes	nr	0.35	2.45	0.00	0.37	2.82

Marley Double Roman Premium granule finish tiles size 420 x 330mm, battens size 38 x 25mm, type 1F reinforced underlay

75mm lap, pitch 30 to 44 degrees	m2	0.82	5.74	9.23	2.25	17.22

Extra for

nailing every tile with aluminium nails	m2	0.50	3.50	0.21	0.56	4.27
verge, 150mm wide plain tile undercloak	m	0.20	1.40	1.02	0.36	2.78
interlocking dry verge system	m	0.20	1.40	4.73	0.92	7.05
segmental ridge tiles	m	0.42	2.94	6.33	1.39	10.66

RATES FOR MEASURED WORK

Marley roof tiles (cont'd)	Unit	Labour hours	Net labour (£)	Net material (£)	O'heads /profit (£)	Total (£)
segmental monoridge tiles	m	0.62	4.34	10.06	2.16	16.56
dry ridge system	m	0.50	3.50	6.24	1.46	11.20
valley trough tiles	m	0.62	4.34	6.29	1.59	12.22
segmental hip tiles	m	0.62	4.34	6.29	1.59	12.22
Marley eaves vent system	m	0.40	2.80	10.27	1.96	15.03
ventilated ridge terminal	nr	0.60	4.20	29.77	5.10	39.07
gas vent terminal	nr	0.60	4.20	46.31	7.58	58.09
soil vent terminal	nr	0.60	4.20	28.67	4.93	37.80
cutting	m	0.20	1.40	0.00	0.21	1.61
holes for pipes	nr	0.35	2.45	0.00	0.37	2.82

Marley Modern smooth finish tiles size 420 x 330mm, battens size 38 x 25mm, type 1F reinforced underlay

75mm lap, pitch 22.5 to 44 degrees	m2	0.82	5.74	9.37	2.27	17.38

Extra for

nailing every tile with aluminium nails	m2	0.05	0.35	0.21	0.08	0.64
interlocking dry verge system	m	0.20	1.40	4.73	0.92	7.05
verge, 150mm wide plain tile undercloak	m	0.20	1.40	1.02	0.36	2.78
Modern ridge tiles	m	0.62	4.34	6.87	1.68	12.89
Modern monoridge	m	0.62	4.34	10.06	2.16	16.56
dry ridge system	m	0.50	3.50	6.24	1.46	11.20
Modern hip tiles	m	0.62	4.34	6.87	1.68	12.89
Marley eaves vent system	m	0.40	2.80	10.48	1.99	15.27
ventilated ridge terminal	nr	0.60	4.20	29.77	5.10	39.07
gas vent terminal	nr	0.60	4.20	46.31	7.58	58.09
soil vent terminal	nr	0.60	4.20	28.67	4.93	37.80
cutting	m	0.20	1.40	0.00	0.21	1.61
holes for pipes	nr	0.35	2.45	0.00	0.37	2.82

ROOF TILING

	Unit	Labour hours	Net labour (£)	Net material (£)	O'heads /profit (£)	Total (£)

Marley Modern smooth finish tiles size 420 x 330mm, battens size 38 x 25mm, type 1F reinforced underlay

100mm lap, pitch 17.5 to 44 degrees	m2	0.83	5.81	10.05	2.38	18.24

Extra for

nailing every tile with aluminium nails	m2	0.05	0.35	0.21	0.08	0.64
interlocking dry verge system	m	0.20	1.40	4.73	0.92	7.05
verge, 130mm wide plain tile undercloak	m	0.20	1.40	1.02	0.36	2.78
Modern ridge tiles	m	0.62	4.34	6.87	1.68	12.89
Modern monoridge tiles	m	0.62	4.34	10.06	2.16	16.56
dry ridge system	m	0.50	3.50	6.24	1.46	11.20
Modern hip tiles	m	0.62	4.34	6.87	1.68	12.89
Marley eaves vent system	nr	0.40	2.80	10.27	1.96	15.03
ventilated ridge terminal	nr	0.60	4.20	29.77	5.10	39.07
gas vent terminal	nr	0.60	4.20	46.31	7.58	58.09
soil vent terminal	nr	0.60	4.20	28.67	4.93	37.80
cutting	m	0.20	1.40	0.00	0.21	1.61
holes for pipes	nr	0.35	2.45	0.00	0.37	2.82

Marley Wessex smooth finish tiles size 420 x 330mm, battens size 38 x 25mm, type 1F reinforced underlay

75mm lap, pitch 15 to 44 degrees	m2	0.83	5.81	10.91	2.51	19.23
100mm lap, pitch 15 to 44 degrees	m2	0.84	5.88	11.67	2.63	20.18

Extra for

nailing every tile with aluminium nails	m2	0.05	0.35	0.21	0.08	0.64
interlocking dry verge system	m	0.20	1.40	4.73	0.92	7.05
verge, 150mm wide plain tile undercloak	m	0.20	1.40	1.02	0.36	2.78
Modern ridge tiles	m	0.62	4.34	6.37	1.61	12.32

RATES FOR MEASURED WORK

Marley roof tiles (cont'd)	Unit	Labour hours	Net labour (£)	Net material (£)	O'heads /profit (£)	Total (£)
Modern monoridge	m	0.62	4.34	10.06	2.16	16.56
dry ridge system	m	0.50	3.50	6.24	1.46	11.20
Modern hip tiles	m	0.62	4.34	6.87	1.68	12.89
Marley eaves vent system	m	0.40	2.80	10.27	1.96	15.03
ventilated ridge terminal	nr	0.60	4.20	29.77	5.10	39.07
gas vent terminal	nr	0.60	4.20	46.31	7.58	58.09
soil vent terminal	nr	0.60	4.20	28.67	4.93	37.80
cutting	m	0.20	1.40	0.00	0.21	1.61
holes for pipes	nr	0.35	2.45	0.00	0.37	2.82

Marley Bold Roll smooth finish tiles size 420 x 330mm, battens size 38 x 25mm, type 1F reinforced underlay

	Unit	Labour hours	Net labour (£)	Net material (£)	O'heads /profit (£)	Total (£)
75mm lap, pitch 17.5 to 44 degrees	m2	0.82	5.74	9.17	2.24	17.15

Extra for

	Unit	Labour hours	Net labour (£)	Net material (£)	O'heads /profit (£)	Total (£)
nailing every tile with aluminium nails	m2	0.05	0.35	0.21	0.08	0.64
interlocking dry verge system	m	0.20	1.40	4.73	0.92	7.05
verge, 150mm wide plain tile undercloak	m	0.20	1.40	1.02	0.36	2.78
segmental ridge tiles	m	0.42	2.94	6.33	1.39	10.66
segmental monoridge tiles	m	0.62	4.34	10.06	2.16	16.56
dry ridge system	m	0.50	3.50	6.24	1.46	11.20
valley trough tiles	m	0.62	4.34	6.29	1.59	12.22
segmental hip tiles	m	0.62	4.34	6.29	1.59	12.22
Marley eaves vent system	m	0.40	2.80	10.27	1.96	15.03
ventilated ridge terminal	nr	0.60	4.20	29.77	5.10	39.07
gas vent terminal	nr	0.60	4.20	46.31	7.58	58.09
soil vent terminal	nr	0.60	4.20	28.67	4.93	37.80
cutting	m	0.20	1.40	0.00	0.21	1.61
holes for pipes	nr	0.35	2.45	0.00	0.37	2.82

ROOF TILING

	Unit	Labour hours	Net labour (£)	Net material (£)	O'heads /profit (£)	Total (£)
Marley Bold Roll granule finish tiles size 420 x 330mm, pitch 30 to 40 degrees, battens size 38 x 25mm, type 1F reinforced underlay						
75mm lap, pitch 30 to 40 degrees	m2	0.83	5.81	9.17	2.25	17.23
100mm lap, pitch 25 to 40 degrees	m2	0.91	6.37	9.83	2.43	18.63
Extra for						
nailing every tile with aluminium nails	m2	0.05	0.35	0.21	0.08	0.64
interlocking dry verge system	m	0.20	1.40	4.73	0.92	7.05
verge, 150mm wide plain tile undercloak	m	0.20	1.40	1.02	0.36	2.78
segmental ridge tiles	m	0.42	2.94	6.33	1.39	10.66
segmental monoridge tiles	m	0.62	4.34	10.06	2.16	16.56
dry ridge system	m	0.50	3.50	6.24	1.46	11.20
valley trough tiles	m	0.62	4.34	6.29	1.59	12.22
segmental hip tiles	m	0.62	4.34	6.29	1.59	12.22
Marley eaves vent system	m	0.40	2.80	10.27	1.96	15.03
ventilated ridge terminal	nr	0.60	4.20	29.77	5.10	39.07
gas vent terminal	nr	0.60	4.20	46.31	7.58	58.09
soil vent terminal	nr	0.60	4.20	28.67	4.93	37.80
cutting	m	0.20	1.40	0.00	0.21	1.61
holes for pipes	nr	0.35	2.45	0.00	0.37	2.82
Marley Bold Roll Premium smooth finish tiles size 420 x 330mm, battens size 38 x 25mm, type 1F reinforced underlay						
75mm lap, pitch 17.5 to 44 degrees	m2	0.82	5.74	10.26	2.40	18.40
Extra for						
nailing every tile with aluminium nails	m2	0.05	0.35	0.21	0.08	0.64
interlocking dry verge system	m	0.20	1.40	4.73	0.92	7.05
verge, 150mm wide plain tile undercloak	m	0.20	1.40	1.02	0.36	2.78
segmental ridge tiles	m	0.42	2.94	6.33	1.39	10.66

RATES FOR MEASURED WORK

Marley roof tiles (cont'd)	Unit	Labour hours	Net labour (£)	Net material (£)	O'heads /profit (£)	Total (£)
segmental monoridge tiles	m	0.62	4.34	10.06	2.16	16.56
dry ridge tiles	m	0.50	3.50	6.24	1.46	11.20
valley trough tiles	m	0.62	4.34	6.29	1.59	12.22
segmental hip tiles	m	0.62	4.34	6.29	1.59	12.22
Marley eaves vent system	m	0.40	2.80	10.27	1.96	15.03
ventilated ridge terminal	nr	0.60	4.20	29.77	5.10	39.07
gas vent terminal	nr	0.60	4.20	46.31	7.58	58.09
soil vent terminal	m	0.60	4.20	28.67	4.93	37.80
cutting	m	0.20	1.40	0.00	0.21	1.61
holes for pipes	nr	0.35	2.45	0.00	0.37	2.82

Redland Renown granular faced or through coloured tiles size 418 x 330mm, 75mm lap, 343mm gauge, pitch 30 to 40 degrees, type 1F reinforced underlay

	Unit					
Battens size 38 x 22mm	m2	0.70	4.90	8.35	1.99	15.24
Battens size 38 x 25mm	m2	0.74	5.18	8.41	2.04	15.63

Extra for

nailing every tile with two aluminium nails	m2	0.09	0.63	0.41	0.16	1.20
cloaked verge tile	m	0.20	1.40	7.17	1.29	9.86
half round ridge or hip tile	m	0.55	3.85	7.17	1.65	12.67
third round ridge tile	nr	0.55	3.85	4.96	1.32	10.13
gas flue ridge terminal	nr	0.70	4.90	46.34	7.69	58.93
Dryvent ridge	m	0.75	5.25	11.32	2.49	19.06
third round hip tile	m	0.55	3.85	5.26	1.37	10.48
Universal valley trough	m	0.30	2.10	12.35	2.17	16.62
cutting	m	0.20	1.40	0.00	0.21	1.61
holes for pipes	nr	0.35	2.45	0.00	0.37	2.82

ROOF TILING

	Unit	Labour hours	Net labour (£)	Net material (£)	O'heads /profit (£)	Total (£)

Redland Renown granular faced or through coloured tiles size 418 x 330mm, 75mm lap, 343mm gauge, pitch 30 to 40 degrees, type 1F reinforced underlay

Battens size 38 x 22mm	m2	0.70	4.90	8.35	1.99	15.24
Battens size 38 x 25mm	m2	0.74	5.18	8.41	2.04	15.63

Extra for

nailing every tile with two aluminium nails	m2	0.09	0.63	0.41	0.16	1.20
cloaked verge tile	m	0.20	1.40	7.17	1.29	9.86
half round ridge or hip tile	m	0.55	3.85	7.17	1.65	12.67
third round ridge tile	nr	0.55	3.85	4.96	1.32	10.13
gas flue ridge terminal	nr	0.70	4.90	46.34	7.69	58.93
Dryvent ridge	m	0.75	5.25	11.32	2.49	19.06
third round hip tile	m	0.55	3.85	5.26	1.37	10.48
Universal valley trough	m	0.30	2.10	12.35	2.17	16.62
cutting	m	0.20	1.40	0.00	0.21	1.61
holes for pipes	nr	0.35	2.45	0.00	0.37	2.82

Redland 50 Double Roman granular faced or through coloured tiles size 418 x 330mm, 75mm lap, 343mm gauge, pitch 30 to 40 degrees, type 1F reinforced underlay

Battens size 38 x 22mm	m2	0.70	4.90	8.35	1.99	15.24
Battens size 38 x 25mm	m2	0.74	5.18	8.41	2.04	15.63

Extra for
nailing every tile with two

aluminium nails	m2	0.09	0.63	0.41	0.16	1.20
cloaked verge tile	m	0.20	1.40	7.17	1.29	9.86
half round ridge or hip tile	m	0.55	3.85	4.96	1.32	10.13
Dryvent ridge	m	0.75	5.25	11.32	2.49	19.06
gas flue ridge terminal	nr	0.70	4.90	46.34	7.69	58.93
third round hip tile	m	0.55	3.85	5.02	1.33	10.20
third round hip and dentil slips 83mm wide	m	0.60	4.20	5.26	1.42	10.88

RATES FOR MEASURED WORK

Redland roof tiles (cont'd)	Unit	Labour hours	Net labour (£)	Net material (£)	O'heads /profit (£)	Total (£)
Universal valley trough	m	0.30	2.10	12.35	2.17	16.62
cutting	m	0.20	1.40	0.00	0.21	1.61
holes for pipes	nr	0.35	2.45	0.00	0.37	2.82
Redland Regent granular faced tiles size 418 x 332mm, 75mm lap, 343mm gauge, pitch 30 to 40 degrees, type 1F reinforced underlay						
Battens size 38 x 22mm	m2	0.77	5.39	9.18	2.19	16.76
Battens size 38 x 25mm	m2	0.82	5.74	9.24	2.25	17.23
Extra for						
nailing every tile with two aluminium nails	m2	0.09	0.63	0.41	0.16	1.20
cloaked verge tile	m	0.20	1.40	7.17	1.29	9.86
half round ridge or hip tile	m	0.55	3.85	4.96	1.32	10.13
Dryvent ridge	m	0.75	5.25	11.32	2.49	19.06
gas flue ridge terminal	nr	0.70	4.90	46.34	7.69	58.93
third round hip tile	m	0.55	3.85	5.02	1.33	10.20
third round hip and dentil slips 41mm wide	m	0.60	4.20	5.26	1.42	10.88
Universal valley trough	m	0.30	2.10	12.35	2.17	16.62
cutting	m	0.20	1.40	0.00	0.21	1.61
holes for pipes	nr	0.35	2.45	0.00	0.37	2.82

ROOF TILING

	Unit	Labour hours	Net labour (£)	Net material (£)	O'heads /profit (£)	Total (£)
Redland Regent through coloured tiles size 418 x 332mm, 100mm lap, 343mm gauge, pitch 17.5 to 22.5 degrees, type 1F reinforced underlay						
Battens size 38 x 22mm	m2	0.70	4.90	9.18	2.11	16.19
Battens size 38 x 25mm	m2	0.74	5.18	9.24	2.16	16.58
Extra for						
nailing every tile with two aluminium nails	m2	0.09	0.63	0.41	0.16	1.20
cloaked verge tile	m	0.20	1.40	7.17	1.29	9.86
half round ridge or hip tile	m	0.55	3.85	4.96	1.32	10.13
Dryvent ridge	m	0.75	5.25	11.32	2.49	19.06
gas flue ridge terminal	nr	0.70	4.90	46.34	7.69	58.93
third round hip tile	m	0.55	3.85	5.02	1.33	10.20
third round hip and dentil slips 41mm wide	m	0.60	4.20	5.26	1.42	10.88
Universal valley trough	m	0.30	2.10	12.35	2.17	16.62
cutting	m	0.20	1.40	0.00	0.21	1.61
holes for pipes	nr	0.35	2.45	0.00	0.37	2.82
Redland Regent through coloured tiles size 418 x 332mm, 75mm lap, 343mm gauge, pitch 22.5 to 40 degrees, type 1F reinforced underlay						
Battens size 38 x 22mm	m2	0.73	5.11	9.18	2.14	16.43
Battens size 38 x 25mm	m2	0.78	5.46	9.24	2.21	16.91
Extra for						
nailing every tile with two aluminium nails	m2	0.09	0.63	0.41	0.16	1.20
cloaked verge tile	m	0.20	1.40	7.17	1.29	9.86
half round ridge or hip tile	m	0.55	3.85	4.96	1.32	10.13
Dryvent ridge	m	0.75	5.25	11.32	2.49	19.06
gas flue ridge terminal	nr	0.70	4.90	46.34	7.69	58.93
third round hip tile	m	0.55	3.85	5.02	1.33	10.20

RATES FOR MEASURED WORK

Redland roof tiles (cont'd)	Unit	Labour hours	Net labour (£)	Net material (£)	O'heads /profit (£)	Total (£)
third round hip and dentil slips 41mm wide	m	0.60	4.20	5.26	1.42	10.88
Universal valley trough	m	0.30	2.10	12.35	2.17	16.62
cutting	m	0.20	1.40	0.00	0.21	1.61
holes for pipes	m	0.35	2.45	0.00	0.37	2.82

Redland Grovebury granular faced tiles size 418 x 322mm, 75mm lap, 343mm gauge, pitch 30 to 44 degrees, type 1F reinforced underlay

Battens size 38 x 22mm	m2	0.75	5.25	9.18	2.16	16.59
Battens size 38 x 25mm	m2	0.78	5.46	9.24	2.21	16.91

Extra for

nailing every tile with two aluminium nails	m2	0.09	0.63	0.41	0.16	1.20
cloaked verge tile	m	0.20	1.40	7.17	1.29	9.86
half round ridge or hip tile	m	0.55	3.85	4.96	1.32	10.13
Dryvent ridge	m	0.75	5.25	11.32	2.49	19.06
gas flue ridge terminal	nr	0.70	4.90	46.34	7.69	58.93
third round hip tile	m	0.55	3.85	5.02	1.33	10.20
third round hip and dentil slips 35mm wide	m	0.60	4.20	5.26	1.42	10.88
Universal valley trough	m	0.30	2.10	12.35	2.17	16.62
cutting	m	0.20	1.40	0.00	0.21	1.61
holes for pipes	nr	0.35	2.45	0.00	0.37	2.82

Redland Grovebury through coloured tiles size 418 x 332mm, 75mm lap, 343mm gauge, pitch 22.5 to 44 degrees, type 1F reinforced underlay

Battens size 38 x 22mm	m2	0.72	5.04	9.18	2.13	16.35
Battens size 38 x 25mm	m2	0.74	5.18	9.24	2.16	16.58

ROOF TILING

	Unit	Labour hours	Net labour (£)	Net material (£)	O'heads /profit (£)	Total (£)
Extra for						
nailing every tile with two aluminium nails	m2	0.09	0.63	0.41	0.16	1.20
cloaked verge tile	m	0.20	1.40	7.17	1.29	9.86
half round ridge or hip tile	m	0.55	3.85	4.96	1.32	10.13
Dryvent ridge	m	0.75	5.25	11.32	2.49	19.06
gas flue ridge terminal	nr	0.70	4.90	46.34	7.69	58.93
third round hip tile	m	0.55	3.85	5.02	1.33	10.20
third round hip and dentil slips 55mm wide	m	0.60	4.20	5.26	1.42	10.88
Universal valley trough	m	0.30	2.10	12.35	2.17	16.62
cutting	m	0.20	1.40	0.00	0.21	1.61
holes for pipes	nr	0.35	2.45	0.00	0.37	2.82

Redland Norfolk pantile through coloured tiles size 381 x 227mm, 100mm lap, 306mm gauge, pitch 22.5 to 25 degrees, type 1F reinforced underlay

Battens size 38 x 22mm	m2	0.80	5.60	9.80	2.31	17.71
Battens size 38 x 25mm	m2	0.89	6.23	9.87	2.42	18.52

Extra for

	Unit					
nailing every tile with two aluminium nails	m2	0.16	1.12	0.68	0.27	2.07
half round ridge or hip tile	m	0.55	3.85	4.96	1.32	10.13
Dryvent ridge	m	0.75	5.25	11.32	2.49	19.06
gas flue ridge terminal	nr	0.70	4.90	46.34	7.69	58.93
third round hip tile	m	0.55	3.85	5.02	1.33	10.20
third round hip and dentil slips 83mm wide	m	0.60	4.20	5.26	1.42	10.88
Universal valley trough	m	0.30	2.10	12.35	2.17	16.62
cutting	m	0.20	1.40	0.00	0.21	1.61
holes for pipes	nr	0.35	2.45	0.00	0.37	2.82

RATES FOR MEASURED WORK

Redland roof tiles (cont'd)	Unit	Labour hours	Net labour (£)	Net material (£)	O'heads /profit (£)	Total (£)
Redland Norfolk pantile through coloured tiles size 381 x 227mm, 75mm lap, 306mm gauge, pitch 25 to 40 degrees, type 1F reinforced underlay						
Battens size 38 x 22mm	m2	0.84	5.88	9.80	2.35	18.03
Battens size 38 x 25mm	m2	0.89	6.23	9.87	2.42	18.52
Extra for						
nailing every tile with two aluminium nails	m2	0.16	1.12	0.68	0.27	2.07
half round ridge or hip tile	m	0.55	0.00	4.96	0.74	5.70
Dryvent ridge	m	0.75	5.25	11.32	2.49	19.06
gas flue ridge terminal	nr	0.70	4.90	46.34	7.69	58.93
third round hip tile	m	0.55	3.85	5.02	1.33	10.20
third round hip and dentil slips 83mm wide	m	0.60	4.20	5.26	1.42	10.88
Universal valley trough	m	0.30	2.10	12.35	2.17	16.62
cutting	m	0.20	1.40	0.00	0.21	1.61
holes for pipes	nr	0.35	2.45	0.00	0.37	2.82
Redland 49 granular faced tiles size 381 x 227mm, 75mm lap, 306mm gauge, pitch 30 to 40 degrees, type 1F reinforced underlay						
Battens size 38 x 22mm	m2	0.80	5.60	8.15	2.06	15.81
Battens size 38 x 25mm	m2	0.84	5.88	8.22	2.12	16.22
Extra for						
nailing every tile with two aluminium nails	m2	0.16	1.12	0.68	0.27	2.07
Dry verge system	m	0.40	2.80	8.27	1.66	12.73
half round ridge or hip tile	m	0.55	3.85	4.96	1.32	10.13
Dryvent ridge	m	0.75	5.25	11.32	2.49	19.06
gas flue ridge terminal	nr	0.70	4.90	46.34	7.69	58.93
third round hip tile	m	0.55	3.85	5.02	1.33	10.20

ROOF TILING

	Unit	Labour hours	Net labour (£)	Net material (£)	O'heads /profit (£)	Total (£)
Universal valley trough	m	0.30	2.10	12.35	2.17	16.62
cutting	m	0.20	1.40	0.00	0.21	1.61
holes for pipes	nr	0.35	2.45	0.00	0.37	2.82

Redland 49 through coloured tiles size 381 x 227mm, 100mm lap, 306mm gauge, pitch 22.5 to 25 degrees, type 1F reinforced underlay

| Battens size 38 x 22mm | m2 | 0.84 | 5.88 | 8.15 | 2.10 | 16.13 |
| Battens size 38 x 25mm | m2 | 0.89 | 6.23 | 8.22 | 2.17 | 16.62 |

Extra for

nailing every tile with two aluminium nails	m2	0.16	1.12	0.68	0.27	2.07
Dry verge system	m	0.40	2.80	8.27	1.66	12.73
half round ridge or hip tile	m	0.55	3.85	4.96	1.32	10.13
Dryvent ridge	m	0.75	5.25	11.32	2.49	19.06
gas flue ridge terminal	nr	0.70	4.90	46.34	7.69	58.93
third round hip tile	m	0.55	3.85	5.02	1.33	10.20
Universal valley trough	m	0.30	2.10	12.35	2.17	16.62
cutting	m	0.20	1.40	0.00	0.21	1.61
holes for pipes	nr	0.35	2.45	0.00	0.37	2.82

Redland 49 through coloured tiles size 381 x 227mm, 75mm lap, 306mm gauge, pitch 25 to 40 degrees, type 1F reinforced underlay

| Battens size 38 x 22mm | m2 | 0.80 | 5.60 | 8.15 | 2.06 | 15.81 |
| Battens size 38 x 25mm | m2 | 0.84 | 5.88 | 8.22 | 2.12 | 16.22 |

Extra for

nailing every tile with two aluminium nails	m2	0.16	1.12	0.68	0.27	2.07
Dry verge system	m	0.40	2.80	8.27	1.66	12.73
half round ridge or hip tile	m	0.55	3.85	4.96	1.32	10.13

RATES FOR MEASURED WORK

Redland roof tiles (cont'd)	Unit	Labour hours	Net labour (£)	Net material (£)	O'heads /profit (£)	Total (£)
Dryvent ridge	m	0.75	5.25	11.32	2.49	19.06
gas flue ridge terminal	nr	0.70	4.90	46.34	7.69	58.93
third round hip tile	m	0.55	3.85	5.02	1.33	10.20
Universal valley trough	m	0.30	2.10	12.35	2.17	16.62
cutting	m	0.20	1.40	0.00	0.21	1.61
holes for pipes	nr	0.35	2.45	0.00	0.37	2.82

Redland Delta through coloured tiles size 430 x 380mm, 75mm lap, 355mm gauge, pitch 17.5 to 44 degrees, type 1F reinforced underlay

Battens size 38 x 22mm	m2	0.67	4.69	10.94	2.34	17.97
Battens size 38 x 25mm	m2	0.70	4.90	11.00	2.38	18.28

Extra for

nailing every tile with two aluminium nails	m2	0.08	0.56	0.34	0.14	1.04
Dry verge system	m	0.40	2.80	8.27	1.66	12.73
Universal Delta ridge	m	0.55	3.85	6.77	1.59	12.21
Delta gas flue ridge	nr	0.75	5.25	57.58	9.42	72.25
Universal angle ridge as hip	m	0.60	4.20	5.78	1.50	11.48
cutting	m	0.20	1.40	0.00	0.21	1.61
holes for pipes	nr	0.35	2.45	0.00	0.37	2.82

Redland Stonewold MK1 through coloured interlocking slates size 430 x 380mm, 75mm lap, 355mm gauge, pitch 17.5 to 44 degrees, type 1F reinforced underlay

Battens size 38 x 22mm	m2	0.67	4.69	10.08	2.22	16.99
Battens size 38 x 25mm	m2	0.70	4.90	10.14	2.26	17.30

Extra for

nailing every slate with two aluminium nails	m2	0.08	0.56	0.34	0.14	1.04
Dry verge system	m	0.40	2.80	8.27	1.66	12.73

ROOF TILING

	Unit	Labour hours	Net labour (£)	Net material (£)	O'heads /profit (£)	Total (£)
Dryvent ridge	m	0.75	5.25	11.42	2.50	19.17
Universal angle ridge	m	0.55	3.85	5.78	1.44	11.07
gas flue ridge terminal	nr	0.70	4.90	46.34	7.69	58.93
Universal angle ridge tile as hip	m	0.60	4.20	5.78	1.50	11.48
cutting	m	0.20	1.40	0.00	0.21	1.61
holes for pipes	nr	0.35	2.45	0.00	0.37	2.82

Redland Stonewold MK2 through coloured interlocking slates size 430 x 380mm, 75mm lap, 355mm gauge, pitch 17.5 to 44 degrees, type 1F reinforced underlay

| Battens size 38 x 22mm | m2 | 0.66 | 4.62 | 9.29 | 2.09 | 16.00 |
| Battens size 38 x 25mm | m2 | 0.70 | 4.90 | 9.36 | 2.14 | 16.40 |

Extra for

nailing every slate with two aluminium nails	m2	0.08	0.56	0.34	0.14	1.04
Dry verge system	m	0.40	2.80	8.27	1.66	12.73
Dryvent ridge	m	0.75	5.25	11.32	2.49	19.06
Universal angle ridge	m	0.55	3.85	5.78	1.44	11.07
gas flue ridge terminal	nr	0.70	4.90	46.34	7.69	58.93
Universal angle ridge tile as hip	m	0.60	4.20	5.78	1.50	11.48
cutting	m	0.20	1.40	0.00	0.21	1.61
holes for pipes	nr	0.35	2.45	0.00	0.37	2.82

Redland Richmond through coloured interlocking slates size 412 x 322mm, 112mm lap, 250mm gauge, pitch 22.5 to 44 degrees, type 1F reinforced underlay

| Battens size 38 x 22mm | m2 | 0.71 | 4.97 | 10.52 | 2.32 | 17.81 |
| Battens size 38 x 25mm | m2 | 0.74 | 5.18 | 10.59 | 2.37 | 18.14 |

Extra for

RATES FOR MEASURED WORK

Redland roof tiles (cont'd)	Unit	Labour hours	Net labour (£)	Net material (£)	O'heads /profit (£)	Total (£)
nailing every slate with two aluminium nails	m2	0.11	0.77	0.46	0.18	1.41
Dry verge system	m	0.40	2.80	8.27	1.66	12.73
Dryvent ridge	m	0.75	5.25	11.32	2.49	19.06
Universal angle ridge	m	0.55	3.85	5.78	1.44	11.07
gas flue ridge terminal	nr	0.70	4.90	46.34	7.69	58.93
Universal angle ridge tile as hip	m	0.60	4.20	5.78	1.50	11.48
cutting	m	0.20	1.40	0.00	0.21	1.61
holes for pipes	nr	0.35	2.45	0.00	0.37	2.82

Redland Plain granular faced or through coloured tiles size 265 x 165mm, 65mm lap, 100mm gauge, 35 degrees pitch, type 1F reinforced underlay

Battens size 32 x 19mm	m2	1.73	12.11	19.98	4.81	36.90
Battens size 32 x 25mm	m2	1.83	12.81	20.19	4.95	37.95

Extra for

nailing every tile with two aluminium nails	m2	0.60	4.20	2.52	1.01	7.73
Dry verge system	m	0.40	2.80	8.27	1.66	12.73
half round ridge tile	m	0.55	3.85	4.96	1.32	10.13
Dryvent ridge	m	0.75	5.25	11.32	2.49	19.06
bonnet hip tile 50 degrees	m	1.00	7.00	17.96	3.74	28.70
third round hip tile	m	0.55	3.85	5.02	1.33	10.20
valley tile 50 degrees	m	1.00	7.00	20.37	4.11	31.48
cutting	m	0.20	1.40	0.00	0.21	1.61
holes for pipes	nr	0.35	2.45	0.00	0.37	2.82

Redland Ornamental granular or through coloured faced tiles size 265 x 165mm, 35mm lap, 115mm gauge 70 degrees pitch to vertical, type 1F reinforced underlay

Battens size 32 x 19mm	m2	1.58	11.06	24.50	5.33	40.89
Battens size 32 x 25mm	m2	1.67	11.69	24.68	5.46	41.83

ROOF TILING

	Unit	Labour hours	Net labour (£)	Net material (£)	O'heads /profit (£)	Total (£)
Extra for						
nailing every tile with two aluminium nails	m2	0.60	4.20	2.52	1.01	7.73
half round ridge tile	m	0.55	3.85	4.96	1.32	10.13
Dryvent ridge	m	0.75	5.25	11.32	2.49	19.06
bonnet hip tile 50 degrees	m	1.00	7.00	17.96	3.74	28.70
third round hip tile	m	0.55	3.85	5.02	1.33	10.20
valley tile 50 degrees	m	1.00	7.00	20.37	4.11	31.48
cutting	m	0.20	1.40	0.00	0.21	1.61
holes for pipes	nr	0.35	2.45	0.00	0.37	2.82
Redland Downland granular faced plain tiles size 265 x 165mm, 65mm lap, 100mm gauge, 35 degrees pitch, type 1F reinforced underlay						
Battens size 32 x 19mm	m2	1.73	12.11	20.06	4.83	37.00
Battens size 32 x 25mm	m2	1.83	12.81	20.27	4.96	38.04
Extra for						
nailing every tile with two aluminium nails	m2	0.60	4.20	2.52	1.01	7.73
half round ridge tile	m	0.55	3.85	4.96	1.32	10.13
Dryvent ridge	m	0.75	5.25	17.96	3.48	26.69
Downland bonnet hip tile 50 degrees	m	1.00	7.00	20.37	4.11	31.48
Downland valley tile 50 degrees	m	1.00	7.00	20.37	4.11	31.48
cutting	m	0.20	1.40	0.00	0.21	1.61
holes for pipes	nr	0.35	2.45	0.00	0.37	2.82
Redland Rosemary red smooth faced tiles size 265 x 165mm, 65mm lap, 100mm gauge, pitch 40 degrees to vertical, type 1F reinforced underlay						
Battens size 38 x 19mm	m2	1.90	13.30	25.35	5.80	44.45
Battens size 38 x 25mm	m2	2.01	14.07	25.56	5.94	45.57

RATES FOR MEASURED WORK

Redland roof tiles (cont'd)	Unit	Labour hours	Net labour (£)	Net material (£)	O'heads /profit (£)	Total (£)
Extra for						
nailing every tile with two aluminium nails	m2	0.60	4.20	2.52	1.01	7.73
half round ridge tile	m	0.55	3.85	4.96	1.32	10.13
bonnet hip tile 50 degrees	m	1.00	7.00	17.96	3.74	28.70
third round hip tile	m	0.55	3.85	5.02	1.33	10.20
valley tile 50 degrees	m	1.00	7.00	17.96	3.74	28.70
cutting	m	0.20	1.40	0.00	0.21	1.61
holes for cutting	nr	0.35	2.45	0.00	0.37	2.82

Redland Rosemary brindle sand tiles 265 x 165mm, 65mm lap, 100mm gauge, 40 degrees to the vertical pitch, type 1F reinforced underlay

Battens size 32 x 19mm	m2	1.73	12.11	30.98	6.46	49.55
Battens size 32 x 25mm	m2	1.83	12.81	31.19	6.60	50.60
Extra for						
nailing every tile with two aluminium nails	m2	0.60	4.20	2.52	1.01	7.73
half round ridge tile	m	0.55	3.85	4.96	1.32	10.13
bonnet hip tile 50 degrees	m	1.00	7.00	17.96	3.74	28.70
third round hip tile	m	0.55	3.85	5.02	1.33	10.20
valley tile 50 degrees	m	1.00	7.00	17.96	3.74	28.70
cutting	m	0.20	1.40	0.00	0.21	1.61
holes for pipes	nr	0.35	2.45	0.00	0.37	2.82

Redland Rosemary Cheslyn sand faced tiles size 265 x 165mm, 65mm lap, 100mm gauge, 40 degrees to vertical pitch, type 1F reinforced underlay

Battens size 32 x 19mm	m2	1.58	11.06	38.59	7.45	57.10
Battens size 32 x 25mm	m2	1.67	11.69	38.90	7.59	58.18

ROOF TILING

	Unit	Labour hours	Net labour (£)	Net material (£)	O'heads /profit (£)	Total (£)
Extra for						
nailing every tile with two aluminium nails	m2	0.60	4.20	2.52	1.01	7.73
half round ridge tile	m	0.55	3.85	4.96	1.32	10.13
bonnet hip tile 50 degrees	m	1.00	7.00	17.96	3.74	28.70
third round hip tile	m	0.55	3.85	5.02	1.33	10.20
valley tile 50 degrees	m	1.00	7.00	17.96	3.74	28.70
cutting	m	0.20	1.40	0.00	0.21	1.61
holes for pipes	nr	0.35	2.45	0.00	0.37	2.82

UNDERFELT AND BATTENS

Treated softwood counter battens nailed with galvanized nails to softwood joists, size

32 x 19mm

450mm centres	m2	0.07	0.49	0.53	0.15	1.17
600mm centres	m2	0.05	0.35	0.32	0.10	0.77
750mm centres	m2	0.03	0.21	0.21	0.06	0.48

32 x 25mm

450mm centres	m2	0.08	0.56	0.68	0.19	1.43
600mm centres	m2	0.06	0.42	0.41	0.12	0.95
750mm centres	m2	0.04	0.28	0.27	0.08	0.63

38 x 22mm

450mm centres	m2	0.10	0.70	0.61	0.20	1.51
600mm centres	m2	0.08	0.56	0.37	0.14	1.07
750mm centres	m2	0.06	0.42	0.24	0.10	0.76

38 x 25mm

450mm centres	m2	0.12	0.84	0.76	0.24	1.84
600mm centres	m2	0.10	0.70	0.45	0.17	1.32
750mm centres	m2	0.08	0.56	0.30	0.13	0.99

50 x 25mm

450mm centres	m2	0.14	0.98	0.99	0.30	2.27

RATES FOR MEASURED WORK

Underfelt and battens (cont'd)	Unit	Labour hours	Net labour (£)	Net material (£)	O'heads /profit (£)	Total (£)
600mm centres	m2	0.12	0.84	0.60	0.22	1.66
750mm centres	m2	0.10	0.70	0.40	0.16	1.26
Slaters type 1F reinforced underlay with 150mm laps secured with batten	15m2	0.03	0.21	0.96	0.18	1.35

FIBRE CEMENT SLATING

	Unit	Labour hours	Net labour (£)	Net material (£)	O'heads /profit (£)	Total (£)

H61 FIBRE CEMENT SLATING

'Duracem' non-asbestos fibre cement slates size 500 x 250mm, pitch over 40 degrees, 38 x 25mm softwood battens, type 1F reinforced underlay

lap 70mm, gauge 215mm	m2	1.06	7.42	19.43	4.03	30.88

'Duracem' non-asbestos fibre cement slates size 500 x 250mm, pitch over 27.5 degrees, 38 x 25mm softwood battens, type 1F reinforced underlay

lap 80mm, gauge 210mm	m2	1.06	7.42	19.05	3.97	30.44

'Duracem' non-asbestos fibre cement slates size 500 x 250mm, pitch over 25 degrees, 38 x 25mm softwood battens, type 1F reinforced underlay

lap 90mm, gauge 205mm	m2	1.07	7.49	18.57	3.91	29.97

'Duracem' non-asbestos fibre cement slates size 500 x 250mm, pitch over 22.5 degrees, 38 x 25mm softwood battens, type 1F reinforced underlay

lap 100mm, gauge 200mm	m2	1.08	7.56	20.36	4.19	32.11

'Duracem' non-asbestos fibre cement slates size 500 x 250mm, pitch 25 to 30 degrees, 38 x 25mm softwood battens, type 1F reinforced underlay

lap 106mm, gauge 197mm	m2	1.09	7.63	19.90	4.13	31.66

RATES FOR MEASURED WORK

Fibre cement slating (cont'd)	Unit	Labour hours	Net labour (£)	Net material (£)	O'heads /profit (£)	Total (£)
'Duracem' non-asbestos fibre cement slates size 600 x 300mm, pitch over 35 degrees, 38 x 25mm softwood battens, type 1F reinforced underlay						
lap 70mm, gauge 265mm	m2	0.89	6.23	17.20	3.51	26.94
'Duracem' non-asbestos fibre cement slates size 600 x 300mm, pitch over 30 degrees, 38 x 25mm softwood battens, type 1F reinforced underlay						
lap 80mm, gauge 260mm	m2	0.90	6.30	16.05	3.35	25.70
'Duracem' non-asbestos fibre cement slates size 600 x 300mm, pitch over 25 degrees, 38 x 25mm softwood battens, type 1F reinforced underlay						
lap 90mm, gauge 255mm	m2	0.90	6.30	17.36	3.55	27.21
'Duracem' non-asbestos fibre cement slates size 600 x 300mm, pitch over 20 degrees, 38 x 25mm softwood battens, type 1F reinforced underlay						
lap 100mm, gauge 250mm	m2	0.91	6.37	17.72	3.61	27.70
'Duracem' non-asbestos fibre cement slates size 600 x 300mm, pitch 20 to 25 degrees, 38 x 25mm softwood battens, type 1F reinforced underlay						
lap 106mm, gauge 247mm	m2	0.91	6.37	17.97	3.65	27.99

FIBRE CEMENT SLATING

	Unit	Labour hours	Net labour (£)	Net material (£)	O'heads /profit (£)	Total (£)
'Eternit 2000' non-asbestos fibre cement slates size 400 x 200mm, pitch over 40 degrees, 38 x 25mm softwood battens, type 1F reinforced underlay						
lap 70mm, gauge 165mm	m2	1.25	8.75	22.00	4.61	35.36
'Eternit 2000' non-asbestos fibre cement slates size 400 x 200mm, pitch 30 to 40 degrees, 38 x 25mm softwood battens, type 1F reinforced underlay						
lap 76mm, gauge 162mm	m2	1.26	8.82	22.61	4.71	36.14
'Eternit 2000' non-asbestos fibre cement slates size 400 x 200mm, pitch 40 to 45 degrees, 38 x 25mm softwood battens, type 1F reinforced underlay						
lap 90mm, gauge 155mm	m2	1.28	8.96	23.58	4.88	37.42
'Eternit 2000' non-asbestos fibre cement slates size 500 x 250mm, pitch over 40 degrees, 38 x 25mm softwood battens, type 1F reinforced underlay						
lap 70mm, gauge 215mm	m2	1.06	7.42	18.90	3.95	30.27
'Eternit 2000' non-asbestos fibre cement slates size 500 x 250mm, pitch over 30 degrees, 38 x 25mm softwood battens, type 1F reinforced underlay						
lap 76mm, gauge 212mm	m2	1.06	7.42	19.20	3.99	30.61

RATES FOR MEASURED WORK

Fibre cement slating (cont'd)	Unit	Labour hours	Net labour (£)	Net material (£)	O'heads /profit (£)	Total (£)

'Eternit 2000' non-asbestos fibre cement slates size 500 x 250mm, pitch over 25 degrees, 38 x 25mm softwood battens, type 1F reinforced underlay

| lap 90mm, gauge 205mm | m2 | 1.07 | 7.49 | 19.79 | 4.09 | 31.37 |

'Eternit 2000' non-asbestos fibre cement slates size 500 x 250mm, pitch over 22.5 degrees, 38 x 25mm softwood battens, type 1F reinforced underlay

| lap 100mm, gauge 200mm | m2 | 1.08 | 7.56 | 20.27 | 4.17 | 32.00 |

'Eternit 2000' non-asbestos fibre cement slates size 500 x 250mm, pitch 25 to 30 degrees, 38 x 25mm softwood battens, type 1F reinforced underlay

| lap 110mm, gauge 195mm | m2 | 1.09 | 7.63 | 20.75 | 4.26 | 32.64 |

'Eternit 2000' non-asbestos fibre cement slates size 600 x 300mm, pitch over 20 degrees, 38 x 25mm softwood battens, type 1F reinforced underlay

| lap 100mm, gauge 250mm | m2 | 0.91 | 6.37 | 18.02 | 3.66 | 28.05 |

'Eternit 2000' non-asbestos fibre cement slates size 600 x 300mm, pitch 25 to 30 degrees, 38 x 25mm softwood battens, type 1F reinforced underlay

| lap 110mm, gauge 245mm | m2 | 0.92 | 6.44 | 18.27 | 3.71 | 28.42 |

FIBRE CEMENT SLATING

	Unit	Labour hours	Net labour (£)	Net material (£)	O'heads /profit (£)	Total (£)
'Rivendale' non-asbestos fibre cement slates size 500 x 250mm, pitch over 22.5 degrees, 38 x 25mm softwood battens, type 1F reinforced underlay						
lap 100mm, gauge 200mm	m2	1.08	7.56	22.39	4.49	34.44
'Rivendale' non-asbestos fibre cement slates size 500 x 250mm, pitch 25 to 30 degrees, 38 x 25mm softwood battens, type 1F reinforced underlay						
lap 110mm, gauge 195mm	m2	1.09	7.63	22.92	4.58	35.13
'Rivendale' non-asbestos fibre cement slates size 600 x 300mm, pitch over 20 degrees, 38 x 25mm softwood battens, type 1F reinforced underlay						
lap 100mm, gauge 250mm	m2	0.91	6.37	19.92	3.94	30.23
'Rivendale' non-asbestos fibre cement slates size 600 x 300mm, pitch 20 to 25 degrees, 38 x 25mm softwood battens, type 1F reinforced underlay						
lap 110mm, gauge 245mm	m2	0.92	6.44	20.21	4.00	30.65
'Country' non-asbestos fibre cement slates size 600 x 300mm, pitch 20 to 25 degrees, 38 x 25mm softwood battens, type 1F reinforced underlay						
lap 100mm, gauge 250mm	m2	0.91	6.37	19.92	3.94	30.23

RATES FOR MEASURED WORK

Fibre cement slating (cont'd)	Unit	Labour hours	Net labour (£)	Net material (£)	O'heads /profit (£)	Total (£)
'Country' non-asbestos fibre cement slates size 600 x 300mm, pitch 20 to 25 degrees, 38 x 25mm softwood battens, type 1F reinforced underlay						
lap 110mm, gauge 245mm	m2	0.92	6.44	20.19	3.99	30.62

NATURAL SLATING

	Unit	Labour hours	Net labour (£)	Net material (£)	O'heads /profit (£)	Total (£)

H62 NATURAL SLATING

Blue/grey slates size 405 x 205mm, 75mm lap, 50 x 25mm softwood battens, type 1F reinforced underlay

Sloping	m2	1.64	11.48	39.12	7.59	58.19
Vertical	m2	1.75	12.25	39.12	7.71	59.08
Mansard	m2	1.75	12.25	39.12	7.71	59.08

Extra for

double eaves course	m	0.50	3.50	5.95	1.42	10.87
single verge undercloak course	m	0.72	5.04	6.17	1.68	12.89
angled ridge or hip tiles	m	0.70	4.90	13.93	2.82	21.65
mitred hips, cutting both sides	m	0.70	4.90	14.65	2.93	22.48
cutting	m	0.60	4.20	0.00	0.63	4.83
hole for small pipes	nr	0.40	2.80	0.00	0.42	3.22
fix only lead soakers	nr	0.45	3.15	0.00	0.47	3.62

Blue/grey slates size 405 x 255mm, 75mm lap, 50 x 25mm softwood battens, type 1F reinforced underlay

Sloping	m2	1.35	9.45	41.38	7.62	58.45
Vertical	m2	1.45	10.15	41.38	7.73	59.26
Mansard	m2	1.45	10.15	41.38	7.73	59.26

Extra for

double eaves course	m	0.50	3.50	6.27	1.47	11.24
single verge undercloak course	m	0.72	5.04	8.00	1.96	15.00
angled ridge or hip tiles	m	0.70	4.90	13.93	2.82	21.65
mitred hips, cutting both sides	m	0.70	4.90	14.65	2.93	22.48
cutting	m	0.60	4.20	0.00	0.63	4.83
hole for small pipes	nr	0.40	2.80	0.00	0.42	3.22
fix only lead soakers	nr	0.45	3.15	0.00	0.47	3.62

RATES FOR MEASURED WORK

Natural slating (cont'd)	Unit	Labour hours	Net labour (£)	Net material (£)	O'heads /profit (£)	Total (£)
Blue/grey slates size 405 x 305mm, 75mm lap, 50 x 25mm softwood battens, type 1F reinforced underlay						
Sloping	m2	1.23	8.61	39.78	7.26	55.65
Vertical	m2	1.33	9.31	39.78	7.36	56.45
Mansard	m2	1.33	9.31	39.78	7.36	56.45
Extra for						
double eaves course	m	0.50	3.50	6.35	1.48	11.33
single verge undercloak course	m	0.72	5.04	9.69	2.21	16.94
angled ridge or hip tiles	m	0.70	4.90	13.93	2.82	21.65
mitred hips, cutting both sides	m	0.70	4.90	14.65	2.93	22.48
cutting	m	0.60	4.20	0.00	0.63	4.83
hole for small pipes	nr	0.40	2.80	0.00	0.42	3.22
fix only lead soakers	nr	0.45	3.15	0.00	0.47	3.62
Blue/grey slates size 460 x 230mm, 75mm lap, 50 x 25mm softwood battens, type 1F reinforced underlay						
Sloping	m2	1.35	9.45	43.21	7.90	60.56
Vertical	m2	1.45	10.15	43.21	8.00	61.36
Mansard	m2	1.45	10.15	43.21	8.00	61.36
Extra for						
double eaves course	m	0.50	3.50	6.79	1.54	11.83
single verge undercloak course	m	0.72	5.04	7.54	1.89	14.47
angled ridge or hip tiles	m	0.70	4.90	13.93	2.82	21.65
mitred hips, cutting both sides	m	0.70	4.90	14.65	2.93	22.48
cutting	m	0.60	4.20	0.00	0.63	4.83
hole for small pipes	nr	0.40	2.80	0.00	0.42	3.22
fix only lead soakers	nr	0.45	3.15	0.00	0.47	3.62

NATURAL SLATING

	Unit	Labour hours	Net labour (£)	Net material (£)	O'heads /profit (£)	Total (£)
Blue/grey slates size 460 x 255mm, 75mm lap, 50 x 25mm softwood battens, type 1F reinforced underlay						
Sloping	m2	1.24	8.68	42.57	7.69	58.94
Vertical	m2	1.35	9.45	42.57	7.80	59.82
Mansard	m2	1.35	9.45	42.57	7.80	59.82
Extra for						
double eaves course	m	0.50	3.50	6.36	1.48	11.34
single verge undercloak course	m	0.72	5.04	8.44	2.02	15.50
angled ridge or hip tiles	m	0.07	0.49	14.04	2.18	16.71
mitred hips, cutting both sides	m	0.70	4.90	14.65	2.93	22.48
cutting	m	0.60	4.20	0.00	0.63	4.83
hole for small pipes	nr	0.40	2.80	0.00	0.42	3.22
fix only lead soakers	nr	0.45	3.15	0.00	0.47	3.62
Blue/grey slates size 460 x 305mm, 75mm lap, 50 x 25mm softwood battens, type 1F reinforced underlay						
Sloping	m2	1.18	8.26	43.47	7.76	59.49
Vertical	m2	1.30	9.10	43.47	7.89	60.46
Mansard	m2	1.30	9.10	43.47	7.89	60.46
Extra for						
double eaves course	m	0.50	3.50	7.81	1.70	13.01
single verge undercloak course	m	0.72	5.04	10.36	2.31	17.71
angled ridge or hip tiles	m	0.70	4.90	13.93	2.82	21.65
mitred hips, cutting both sides	m	0.70	4.90	14.65	2.93	22.48
cutting	m	0.60	4.20	0.00	0.63	4.83
hole for small pipes	nr	0.40	2.80	0.00	0.42	3.22
fix only lead soakers	nr	0.45	3.15	0.00	0.47	3.62

RATES FOR MEASURED WORK

Natural slating (cont'd)	Unit	Labour hours	Net labour (£)	Net material (£)	O'heads /profit (£)	Total (£)
Blue/grey slates size 510 x 255mm, 75mm lap, 50 x 25mm softwood battens, type 1F reinforced underlay						
Sloping	m2	1.27	8.89	51.90	9.12	69.91
Vertical	m2	1.37	9.59	51.90	9.22	70.71
Mansard	m2	1.37	9.59	51.90	9.22	70.71
Extra for						
double eaves course	m	0.50	3.50	10.68	2.13	16.31
single verge undercloak course	m	0.72	5.04	10.90	2.39	18.33
angled ridge or hip tiles	m	0.70	4.90	13.93	2.82	21.65
mitred hips, cutting both sides	m	0.70	4.90	14.65	2.93	22.48
cutting	m	0.60	4.20	0.00	0.63	4.83
hole for small pipes	nr	0.40	2.80	0.00	0.42	3.22
fix only lead soakers	nr	0.45	3.15	0.00	0.47	3.62
Blue/grey slates size 510 x 305mm, 75mm lap, 50 x 25mm softwood battens, type 1F reinforced underlay						
Sloping	m2	1.01	7.07	49.05	8.42	64.54
Vertical	m2	1.10	7.70	49.05	8.51	65.26
Mansard	m2	1.10	7.70	49.05	8.51	65.26
Extra for						
double eaves course	m	0.50	3.50	10.16	2.05	15.71
single verge undercloak course	m	0.72	5.04	17.64	3.40	26.08
angled ridge or hip tiles	m	0.70	4.90	13.93	2.82	21.65
mitred hips, cutting both sides	m	0.70	4.90	14.65	2.93	22.48
cutting	m	0.60	4.20	0.00	0.63	4.83
hole for small pipes	nr	0.40	2.80	0.00	0.42	3.22
fix only lead soakers	nr	0.45	3.15	0.00	0.47	3.62

NATURAL SLATING

	Unit	Labour hours	Net labour (£)	Net material (£)	O'heads /profit (£)	Total (£)
Blue/grey slates size 560 x 280mm, 75mm lap, 50 x 25mm softwood battens, type 1F reinforced underlay						
Sloping	m2	1.02	7.14	56.84	9.60	73.58
Extra for						
double eaves course	m	0.50	3.50	13.07	2.49	19.06
single verge undercloak course	m	0.72	5.04	14.64	2.95	22.63
angled ridge or hip tiles	m	0.70	4.90	14.25	2.87	22.02
mitred hips, cutting both sides	m	0.70	4.90	14.65	2.93	22.48
cutting	m	0.60	4.20	0.00	0.63	4.83
hole for small pipes	nr	0.40	2.80	0.00	0.42	3.22
fix only lead soakers	nr	0.45	3.15	0.00	0.47	3.62
Blue/grey slates size 560 x 305mm, 75mm lap, 50 x 25mm softwood battens, type 1F reinforced underlay						
Sloping	m2	0.96	6.72	58.72	9.82	75.26
Extra for						
double eaves course	m	0.50	3.50	13.03	2.48	19.01
single verge undercloak course	m	0.72	5.04	16.04	3.16	24.24
angled ridge or hip tiles	m	0.70	4.90	13.93	2.82	21.65
mitred hips, cutting both sides	m	0.70	4.90	14.65	2.93	22.48
cutting	m	0.60	4.20	0.00	0.63	4.83
hole for small pipes	nr	0.40	2.80	0.00	0.42	3.22
fix only lead soakers	nr	0.45	3.15	0.00	0.47	3.62

RATES FOR MEASURED WORK

	Unit	Labour hours	Net labour (£)	Net material (£)	O'heads /profit (£)	Total (£)
Blue/grey slates size 610 x 305mm, 75mm lap, 50 x 25mm softwood battens, type 1F reinforced underlay						
Sloping	m2	0.89	6.23	68.30	11.18	85.71
Extra for						
double eaves course	m	0.50	3.50	17.71	3.18	24.39
single verge undercloak course	m	0.72	5.04	17.82	3.43	26.29
angled ridge or hip tiles	m	0.70	4.90	13.93	2.82	21.65
mitred hips, cutting both sides	m	0.70	4.90	14.65	2.93	22.48
cutting	m	0.60	4.20	0.00	0.63	4.83
hole for small pipes	nr	0.40	2.80	0.00	0.42	3.22
fix only lead soakers	nr	0.45	3.15	0.00	0.47	3.62
Westmoreland green slates Kirkstone light sea green random length slates, 500-700mm long, thin quality, 75mm lap, diminishing courses eaves to ridge, copper nailed, 50 x 25mm pressure impregnated battens nailed, underlay type 1F felt, clout nailed						
Sloping 30 degrees from horizontal	m2	1.43	10.01	107.32	17.60	134.93
Vertical slating	m2	1.60	11.20	107.32	17.78	136.30
Extra for						
top edge	m	0.40	2.80	2.37	0.78	5.95
eaves courses	m	0.50	3.50	15.62	2.87	21.99
mitred hips, cutting both sides	m	1.40	9.80	7.81	2.64	20.25
square abutments	m	0.60	4.20	2.69	1.03	7.92
valleys and gutters (each side)	m	0.80	5.60	6.25	1.78	13.63
skew and circular cutting	m	1.30	9.10	5.00	2.11	16.21

NATURAL SLATING

	Unit	Labour hours	Net labour (£)	Net material (£)	O'heads /profit (£)	Total (£)
Kirkstone light sea green random lengths roofing slates 300 to 225mm long standard quality peggies; 75mm lap as previous						
sloping 30 degrees from horizontal	m2	1.97	13.79	75.40	13.38	102.57
vertical slating	m2	2.21	15.47	75.40	13.63	104.50
Extra for						
top edge	m	0.60	4.20	2.47	1.00	7.67
eaves course	m	0.75	5.25	17.66	3.44	26.35
mitred hips, cutting both sides	m	2.00	14.00	8.83	3.42	26.25
square abutments	m	0.90	6.30	3.00	1.40	10.70
valley and gutters (each side)	m	1.20	8.40	7.07	2.32	17.79
skew and circular cutting	m	1.90	13.30	6.01	2.90	22.21

RATES FOR MEASURED WORK

	Unit	Labour hours	Net labour (£)	Net material (£)	O'heads /profit (£)	Total (£)
H63 RECONSTRUCTED STONE SLATING						
Marley Monarch interlocking slate 38 x 25mm, type 1F reinforced underlay						
75mm lap, pitch 25 to 90 degrees	m2	1.24	8.68	20.27	4.34	33.29
100mm lap, pitch 25 to 90 degrees	m2	1.26	8.82	21.98	4.62	35.42
Extra for						
nailing every tile with aluminium nails	m2	0.09	0.63	0.21	0.13	0.97
interlocking dry verge system	m	0.20	1.40	4.73	0.92	7.05
Modern ridge tiles	m	0.62	4.34	6.87	1.68	12.89
Modern monoridge	m	0.62	4.34	10.06	2.16	16.56
dry ridge system	m	0.50	3.50	6.24	1.46	11.20
Modern hip tiles	m	0.62	4.34	6.87	1.68	12.89
ventilated ridge terminal	nr	0.60	4.20	29.77	5.10	39.07
gas vent terminal	nr	0.60	4.20	46.31	7.58	58.09
soil vent terminal	nr	0.60	4.20	28.67	4.93	37.80
cutting	m	0.20	1.40	0.00	0.21	1.61
holes for pipes	nr	0.35	2.45	0.00	0.37	2.82
Redland Cambrian through coloured slates interlocking size 300 x 336mm, pitch 25 to 69 degrees, batten size 38 x 25mm, type 1F reinforcing underlay						
50mm lap, 250mm gauge	m2	0.90	6.30	21.37	4.15	31.82
90mm lap, 210mm gauge.	m2	1.00	7.00	25.52	4.88	37.40
Extra for						
nailing every tile with two aluminium nails	m2	0.60	4.20	0.41	0.69	5.30

RECONSTRUCTED STONE SLATING

	Unit	Labour hours	Net labour (£)	Net material (£)	O'heads /profit (£)	Total (£)
half round ridge tile	m	0.55	3.85	4.96	1.32	10.13
Dryvent ridge	m	0.75	5.25	11.32	2.49	19.06
Universal angle ridge	m	0.55	3.85	5.78	1.44	11.07
gas flue ridge terminal	nr	0.70	4.90	46.34	7.69	58.93
Universal angle ridge tile as hip	m	0.60	4.20	5.78	1.50	11.48
cutting	m	0.20	1.40	0.00	0.21	1.61
holes for pipes	nr	0.35	2.45	0.00	0.37	2.82

RATES FOR MEASURED WORK

	Unit	Labour hours	Net labour (£)	Net material (£)	O'heads /profit (£)	Total (£)

H64 TIMBER SHINGLING

Western Red Cedar sawn shingles 400mm long; random width to 38 x 25mm softwood battens with silicon/bronze nails and type 1F reinforced felt underlay

Roof coverings 14 to 22 degrees slope, gauge 95mm	m2	1.20	8.40	22.72	4.67	35.79
Roof coverings 22 to 45 degrees slope, gauge 125mm	m2	1.30	9.10	18.79	4.18	32.07
Extra for						
eaves double course	m	0.10	0.70	5.01	0.86	6.57
ridge and hip capping	m	0.30	2.10	7.51	1.44	11.05
double course at ridge	m	1.10	7.70	5.01	1.91	14.62
verges, underlocking course	m	0.30	2.10	5.01	1.07	8.18
raking cutting	m	0.10	0.70	0.00	0.10	0.80

LEAD SHEET COVERINGS

	Unit	Labour hours	Net labour (£)	Net material (£)	O'heads /profit (£)	Total (£)

H71 LEAD SHEET COVERINGS

Roof coverings, milled sheet lead to BS1178

Flat roofing, pitch less than 10 degrees to the horizontal

code 4	m2	4.00	28.00	24.76	7.91	60.67
code 5	m2	4.20	29.40	30.82	9.03	69.25
code 6	m2	4.40	30.80	36.52	10.10	77.42
code 7	m2	4.60	32.20	43.33	11.33	86.86
code 8	m2	4.80	33.60	48.85	12.37	94.82

Dormers, pitch less than 10 degrees to the horizontal

code 4	m2	4.75	33.25	26.00	8.89	68.14
code 5	m2	5.00	35.00	32.40	10.11	77.51
code 6	m2	5.25	36.75	38.35	11.27	86.37
code 7	m2	5.25	36.75	45.50	12.34	94.59
code 8	m2	5.75	40.25	51.29	13.73	105.27

Sloping roofing, pitch 10 degrees to 50 degrees

code 4	m2	4.20	29.40	24.76	8.12	62.28
code 5	m2	4.40	30.80	30.82	9.24	70.86
code 6	m2	4.60	32.20	36.52	10.31	79.03
code 7	m2	4.80	33.60	43.33	11.54	88.47
code 8	m2	5.00	35.00	48.85	12.58	96.43

Dormers, pitch 10 degrees to 50 degrees

code 4	m2	4.75	33.25	26.00	8.89	68.14
code 5	m2	5.00	35.00	32.40	10.11	77.51
code 6	m2	5.25	36.75	38.35	11.27	86.37
code 7	m2	5.50	38.50	45.50	12.60	96.60
code 8	m2	5.75	40.25	51.29	13.73	105.27

RATES FOR MEASURED WORK

Lead sheet coverings (cont'd)	Unit	Labour hours	Net labour (£)	Net material (£)	O'heads /profit (£)	Total (£)
Sloping, roofing pitch over 50 degrees						
code 4	m2	4.40	30.80	24.76	8.33	63.89
code 5	m2	4.60	32.20	30.82	9.45	72.47
code 6	m2	4.80	33.60	36.52	10.52	80.64
code 7	m2	5.00	35.00	43.33	11.75	90.08
code 8	m2	5.20	36.40	48.85	12.79	98.04
Dormers, pitch over 50 degrees						
code 4	m2	4.75	33.25	26.00	8.89	68.14
code 5	m2	5.00	35.00	32.40	10.11	77.51
code 6	m2	5.25	36.75	38.35	11.27	86.37
code 7	m2	5.50	38.50	45.50	12.60	96.60
code 8	m2	5.75	40.25	51.29	13.73	105.27
Flashings, horizontal, girth 150mm						
code 4	m	0.40	2.80	3.33	0.92	7.05
code 5	m	0.45	3.15	4.86	1.20	9.21
Flashings, horizontal, girth 200mm						
code 4	m	0.55	3.85	5.20	1.36	10.41
code 5	m	0.60	4.20	6.48	1.60	12.28
Flashings, horizontal, girth 300mm						
code 4	m	0.50	3.50	7.80	1.69	12.99
code 5	m	0.55	3.85	9.72	2.04	15.61
Flashings, sloping, girth 150mm						
code 4	m	0.40	2.80	4.28	1.06	8.14
code 5	m	0.45	3.15	5.34	1.27	9.76
Flashings, sloping, girth 200mm						
code 4	m	0.55	3.85	5.72	1.44	11.01
code 5	m	0.60	4.20	7.13	1.70	13.03

LEAD SHEET COVERINGS

	Unit	Labour hours	Net labour (£)	Net material (£)	O'heads /profit (£)	Total (£)
Flashings, sloping, girth 300mm						
code 4	m	0.80	5.60	8.58	2.13	16.31
code 5	m	0.85	5.95	10.70	2.50	19.15
Flashings, stepped, sloping girth 150mm						
code 4	m	0.50	3.50	4.67	1.23	9.40
code 5	m	0.55	3.85	5.85	1.45	11.15
Flashings, stepped, sloping girth 200mm						
code 4	m	0.90	6.30	6.24	1.88	14.42
code 5	m	1.00	7.00	7.77	2.22	16.99
Flashings, stepped, sloping girth 300mm						
code 4	m	1.10	7.70	9.37	2.56	19.63
code 5	m	1.20	8.40	11.67	3.01	23.08
Aprons, horizontal, girth 200mm						
code 4	m	0.55	3.85	4.67	1.28	9.80
code 5	m	1.60	11.20	6.48	2.65	20.33
Aprons, horizontal, girth 300mm						
code 4	m	0.80	5.60	7.80	2.01	15.41
code 5	m	0.90	6.30	9.72	2.40	18.42
Aprons, horizontal, girth 400mm						
code 4	m	1.10	7.70	10.40	2.71	20.81
code 5	m	1.20	8.40	12.96	3.20	24.56
Aprons, sloping, girth 200mm						
code 4	m	0.65	4.55	5.71	1.54	11.80
code 5	m	0.70	4.90	7.13	1.80	13.83

RATES FOR MEASURED WORK

Lead sheet coverings (cont'd)	Unit	Labour hours	Net labour (£)	Net material (£)	O'heads /profit (£)	Total (£)
Aprons, sloping, girth 300mm						
code 4	m	0.90	6.30	8.58	2.23	17.11
code 5	m	1.00	7.00	10.70	2.65	20.35
Aprons, sloping, girth 400mm						
code 4	m	1.20	8.40	11.43	2.97	22.80
code 5	m	1.30	9.10	14.27	3.51	26.88
Sills, horizontal, girth 200mm						
code 4	m	0.55	3.85	5.20	1.36	10.41
code 5	m	0.60	4.20	6.48	1.60	12.28
Sills, horizontal, girth 300mm						
code 4	m	0.80	5.60	7.80	2.01	15.41
code 5	m	0.90	6.30	9.72	2.40	18.42
Sills, horizontal, girth 400mm						
code 4	m	1.10	7.70	10.40	2.71	20.81
code 5	m	1.20	8.40	12.96	3.20	24.56
Cappings, horizontal, girth 200mm						
code 4	m	0.55	3.85	5.20	1.36	10.41
code 5	m	0.60	4.20	6.48	1.60	12.28
Cappings, horizontal, girth 300mm						
code 4	m	0.80	5.60	7.80	2.01	15.41
code 5	m	0.90	6.30	9.72	2.40	18.42
Cappings, horizontal, girth 400mm						
code 4	m	1.10	7.70	10.40	2.71	20.81
code 5	m	1.20	8.40	12.96	3.20	24.56
Hips, sloping, girth 200mm						
code 4	m	0.65	4.55	5.20	1.46	11.21
code 5	m	0.75	5.25	6.48	1.76	13.49

LEAD SHEET COVERINGS

	Unit	Labour hours	Net labour (£)	Net material (£)	O'heads /profit (£)	Total (£)
Hips, sloping, girth 300mm						
code 4	m	0.90	6.30	7.80	2.11	16.21
code 5	m	1.00	7.00	9.72	2.51	19.23
Hips, sloping, girth 400mm						
code 4	m	1.20	8.40	10.40	2.82	21.62
code 5	m	1.30	9.10	12.96	3.31	25.37
Kerbs, horizontal, girth 300mm						
code 4	m	0.80	5.60	7.80	2.01	15.41
code 5	m	0.90	6.30	9.72	2.40	18.42
Kerbs, horizontal, girth 400mm						
code 4	m	1.10	7.70	10.40	2.71	20.81
code 5	m	1.20	8.40	12.96	3.20	24.56
Ridges, horizontal, girth 200mm						
code 4	m	0.55	3.85	5.20	1.36	10.41
code 5	m	0.60	4.20	6.48	1.60	12.28
Ridges, horizontal, girth 300mm						
code 4	m	0.80	5.60	7.80	2.01	15.41
code 5	m	0.90	6.30	9.72	2.40	18.42
Ridges, horizontal, girth 400mm						
code 4	m	1.10	7.70	10.40	2.71	20.81
code 5	m	1.20	8.40	12.96	3.20	24.56
Valleys, sloping, girth 400mm						
code 4	m	1.10	7.70	10.40	2.71	20.81
code 5	m	1.20	8.40	12.96	3.20	24.56
Valleys, sloping, girth 500mm						
code 4	m	1.30	9.10	13.00	3.31	25.41
code 5	m	1.40	9.80	16.20	3.90	29.90

RATES FOR MEASURED WORK

Lead sheet coverings (cont'd)	Unit	Labour hours	Net labour (£)	Net material (£)	O'heads /profit (£)	Total (£)
Valleys, sloping, girth 800mm						
code 4	m	1.50	10.50	20.80	4.70	36.00
code 5	m	1.60	11.20	25.92	5.57	42.69
Gutters, sloping, girth 400mm						
code 5	m	1.20	8.40	12.96	3.20	24.56
code 6	m	1.40	9.80	15.34	3.77	28.91
code 7	m	1.60	11.20	18.20	4.41	33.81
code 8	m	1.70	11.90	20.52	4.86	37.28
Gutters, sloping, girth 600mm						
code 5	m	1.40	9.80	16.20	3.90	29.90
code 6	m	1.60	11.20	19.17	4.56	34.93
code 7	m	1.80	12.60	22.75	5.30	40.65
code 8	m	1.90	13.30	25.64	5.84	44.78
Gutters, sloping, girth 800mm						
code 5	m	1.60	11.20	25.92	5.57	42.69
code 6	m	1.80	12.60	30.68	6.49	49.77
code 7	m	2.00	14.00	36.39	7.56	57.95
code 8	m	2.10	14.70	41.03	8.36	64.09
Edges, welted						
code 5	m	0.35	2.45	0.00	0.37	2.82
code 6	m	0.45	3.15	0.00	0.47	3.62
code 7	m	0.55	3.85	0.00	0.58	4.43
code 8	m	0.60	4.20	0.00	0.63	4.83
Edges, beaded						
code 5	m	0.35	2.45	0.00	0.37	2.82
code 6	m	0.45	3.15	0.00	0.47	3.62
code 7	m	0.55	3.85	0.00	0.58	4.43
code 8	m	0.60	4.20	0.00	0.63	4.83

LEAD SHEET COVERINGS

	Unit	Labour hours	Net labour (£)	Net material (£)	O'heads /profit (£)	Total (£)
Dressing over corrugated sheeting, down corrugations						
code 5	m	0.20	1.40	0.00	0.21	1.61
code 6	m	0.25	1.75	0.00	0.26	2.01
code 7	m	0.30	2.10	0.00	0.32	2.42
code 8	m	0.40	2.80	0.00	0.42	3.22
Dressing over corrugated sheeting, across corrugations						
code 5	m	1.05	7.35	0.00	1.10	8.45
code 6	m	1.15	8.05	0.00	1.21	9.26
code 7	m	1.20	8.40	0.00	1.26	9.66
code 8	m	1.25	8.75	0.00	1.31	10.06
Dressing over slating and tiling						
code 5	m	0.20	1.40	0.00	0.21	1.61
code 6	m	0.25	1.75	0.00	0.26	2.01
code 7	m	0.30	2.10	0.00	0.32	2.42
code 8	m	0.40	2.80	0.00	0.42	3.22
Dressing over glass and glazing bar						
code 5	m	0.20	1.40	0.00	0.21	1.61
code 6	m	0.25	1.75	0.00	0.26	2.01
code 7	m	0.30	2.10	0.00	0.32	2.42
code 8	m	0.40	2.80	0.00	0.42	3.22
Soakers size 150 x 150mm						
code 3	nr	0.25	1.75	0.38	0.32	2.45
code 4	nr	0.30	2.10	0.50	0.39	2.99
Soakers size 200 x 300mm						
code 3	nr	0.30	2.10	0.99	0.46	3.55
code 4	nr	0.35	2.45	1.35	0.57	4.37
Soakers size 300 x 150mm						
code 3	nr	0.35	2.45	0.73	0.48	3.66
code 4	nr	0.40	2.80	1.02	0.57	4.39

RATES FOR MEASURED WORK

Lead sheet coverings (cont'd)	Unit	Labour hours	Net labour (£)	Net material (£)	O'heads /profit (£)	Total (£)
Slates, size 400 x 400mm with 200mm high collar, 100mm diameter						
code 4	nr	1.50	10.50	8.32	2.82	21.64
code 5	nr	1.60	11.20	10.37	3.24	24.81
Slates, size 400 x 400mm with 200mm high collar, 150mm diameter						
code 4	nr	1.70	11.90	9.97	3.28	25.15
code 5	nr	1.80	12.60	12.47	3.76	28.83
Dots, cast lead						
code 3	nr	0.80	5.60	0.79	0.96	7.35
code 4	nr	0.85	5.95	0.89	1.03	7.87
code 5	nr	0.90	6.30	1.00	1.09	8.39
Dots, soldered						
code 3	nr	0.90	6.30	4.83	1.67	12.80
code 4	nr	1.00	7.00	5.15	1.82	13.97
code 5	nr	1.10	7.70	5.35	1.96	15.01

ALUMINIUM SHEET COVERINGS

	Unit	Labour hours	Net labour (£)	Net material (£)	O'heads /profit (£)	Total (£)
H72 ALUMINIUM SHEET COVERINGS						
0.6mm commercial grade aluminium to BS1470						
Roof covering						
flat	m2	3.20	22.40	8.17	4.59	35.16
sloping 10 to 50 degrees	m2	3.50	24.50	8.17	4.90	37.57
sloping or vertical over 50 degrees	m2	4.00	28.00	8.17	5.43	41.60
Flashing						
150mm girth	m	0.50	3.50	1.46	0.74	5.70
200mm girth	m	0.60	4.20	2.66	1.03	7.89
375mm girth	m	0.90	6.30	3.45	1.46	11.21
Stepped flashing						
150mm girth	m	0.60	4.20	1.46	0.85	6.51
200mm girth	m	0.70	4.90	2.66	1.13	8.69
375mm girth	m	1.00	7.00	3.45	1.57	12.02
Apron						
200mm girth	m	0.60	4.20	2.66	1.03	7.89
375mm girth	m	0.90	6.30	3.45	1.46	11.21
450mm girth	m	1.20	8.40	3.96	1.85	14.21
Capping to ridge or hip						
200mm girth	m	0.60	4.20	2.66	1.03	7.89
375mm girth	m	0.90	6.30	3.45	1.46	11.21
450mm girth	m	1.20	8.40	3.96	1.85	14.21
Lining to valley or gutter						
450mm girth	m	1.20	8.40	3.96	1.85	14.21
600mm girth	m	1.60	11.20	5.00	2.43	18.63
900mm girth	m	2.00	14.00	7.35	3.20	24.55
Raking cutting	m	0.15	1.05	0.00	0.16	1.21

RATES FOR MEASURED WORK

Aluminium sheet coverings (cont'd)	Unit	Labour hours	Net labour (£)	Net material (£)	O'heads /profit (£)	Total (£)
Curved cutting	m	0.20	1.40	0.00	0.21	1.61
Welted edge	m	0.25	1.75	0.00	0.26	2.01
Beaded edge	m	0.25	1.75	0.00	0.26	2.01
Wedging into groove with aluminium wedges	m	0.25	1.75	0.23	0.30	2.28

0.8mm commercial grade aluminium BS1470

Roof covering

flat	m2	3.20	22.40	10.49	4.93	37.82
sloping 10 to 50 degrees	m2	3.50	24.50	10.49	5.25	40.24
sloping or vertical over 50 degrees	m2	4.00	28.00	10.49	5.77	44.26

Flashing

150mm girth	m	0.50	3.50	1.88	0.81	6.19
225mm girth	m	0.60	4.20	3.40	1.14	8.74
300mm girth	m	0.90	6.30	4.44	1.61	12.35

Stepped flashing

150mm girth	m	0.60	4.20	1.88	0.91	6.99
225mm girth	m	0.70	4.90	3.40	1.24	9.54
300mm girth	m	1.00	7.00	4.44	1.72	13.16

Apron

225mm girth	m	0.60	4.20	3.40	1.14	8.74
300mm girth	m	0.90	6.30	4.44	1.61	12.35
450mm girth	m	1.20	8.40	5.08	2.02	15.50

Capping to ridge or hip

225mm girth	m	0.60	4.20	3.40	1.14	8.74
300mm girth	m	0.90	6.30	4.44	1.61	12.35
450mm girth	m	1.20	8.40	5.08	2.02	15.50

ALUMINIUM SHEET COVERINGS

	Unit	Labour hours	Net labour (£)	Net material (£)	O'heads /profit (£)	Total (£)
Lining to valley or gutter						
450mm girth	m	1.20	8.40	5.08	2.02	15.50
600mm girth	m	1.60	11.20	6.44	2.65	20.29
900mm girth	m	2.00	14.00	9.44	3.52	26.96
Raking cutting	m	0.15	1.05	0.00	0.16	1.21
Curved cutting	m	0.20	1.40	0.00	0.21	1.61
Welted edge	m	0.25	1.75	0.00	0.26	2.01
Beaded edges	m	0.25	1.75	0.00	0.26	2.01
Wedging into groove with aluminium wedges	m	0.25	1.75	0.13	0.28	2.16

RATES FOR MEASURED WORK

	Unit	Labour hours	Net labour (£)	Net material (£)	O'heads /profit (£)	Total (£)

H73 COPPER SHEET COVERINGS

0.55mm thick copper sheeting to BS2870

Roof coverings

flat	m2	3.20	22.40	25.40	7.17	54.97
sloping 10 to 50 degrees	m2	3.50	24.50	25.40	7.48	57.39
sloping or vertical over 50 degrees	m2	4.00	28.00	25.40	8.01	61.41

Flashing

150mm girth	m	0.50	3.50	3.80	1.09	8.39
200mm girth	m	0.60	4.20	5.08	1.39	10.67
300mm girth	m	0.90	6.30	7.62	2.09	16.01

Stepped flashing

150mm girth	m	0.60	4.20	3.80	1.20	9.20
200mm girth	m	0.70	4.90	5.08	1.50	11.48
300mm girth	m	1.00	7.00	7.62	2.19	16.81

Apron

200mm girth	m	0.60	4.20	5.08	1.39	10.67
300mm girth	m	0.90	6.30	7.62	2.09	16.01
400mm girth	m	1.20	8.40	10.16	2.78	21.34

Capping to ridge or hip

200mm girth	m	0.60	4.20	5.08	1.39	10.67
300mm girth	m	0.90	6.30	7.62	2.09	16.01
400mm girth	m	1.20	8.40	10.16	2.78	21.34

Lining to valley or gutter

400mm girth	m	1.20	8.40	10.16	2.78	21.34
600mm girth	m	1.60	11.20	15.24	3.97	30.41
800mm girth	m	2.00	14.00	20.32	5.15	39.47

Raking cutting	m	0.15	1.05	0.00	0.16	1.21

COPPER SHEET COVERINGS

	Unit	Labour hours	Net labour (£)	Net material (£)	O'heads /profit (£)	Total (£)
Curved cutting	m	0.20	1.40	0.00	0.21	1.61
Welted edge	m	0.25	1.75	0.00	0.26	2.01
Beaded edge	m	0.25	1.75	0.00	0.26	2.01
Wedging into groove with copper wedges	m	0.25	1.75	0.32	0.31	2.38
Brazed angle	m	1.00	7.00	0.00	1.05	8.05
Brazed seam	m	1.00	7.00	0.00	1.05	8.05

0.7mm thick copper sheeting to BS2870

Roof covering

flat	m2	3.20	22.40	32.45	8.23	63.08
sloping 10 to 50 degrees	m2	3.50	24.50	32.45	8.54	65.49
sloping or vertical over 50 degrees	m2	4.00	28.00	32.45	9.07	69.52

Flashing

150mm girth	m	0.50	3.50	4.87	1.26	9.63
200mm girth	m	0.60	4.20	6.49	1.60	12.29
300mm girth	m	0.90	6.30	9.73	2.40	18.43

Stepped flashing

150mm girth	m	0.60	4.20	4.87	1.36	10.43
200mm girth	m	0.70	4.90	6.49	1.71	13.10
300mm girth	m	1.00	7.00	9.73	2.51	19.24

Apron

200mm girth	m	0.50	3.50	6.49	1.50	11.49
300mm girth	m	0.80	5.60	9.73	2.30	17.63
400mm girth	m	1.00	7.00	12.98	3.00	22.98

Capping to ridge or hip

200mm girth	m	0.60	4.20	6.49	1.60	12.29

RATES FOR MEASURED WORK

Copper sheet coverings (cont'd)	Unit	Labour hours	Net labour (£)	Net material (£)	O'heads /profit (£)	Total (£)
300mm girth	m	0.90	6.30	9.73	2.40	18.43
400mm girth	m	1.20	8.40	12.98	3.21	24.59
Lining to valley or gutter						
400mm girth	m	1.20	8.40	12.98	3.21	24.59
600mm girth	m	1.60	11.20	19.47	4.60	35.27
800mm girth	m	2.00	14.00	25.96	5.99	45.95
Raking cutting	m	0.15	1.05	0.00	0.16	1.21
Curved cutting	m	0.20	1.40	0.00	0.21	1.61
Welted edge	m	0.25	1.75	0.00	0.26	2.01
Beaded edge	m	0.25	1.75	0.00	0.26	2.01
Wedging into groove with copper wedges	m	0.25	1.75	0.32	0.31	2.38
Brazed angle	m	1.00	7.00	0.00	1.05	8.05
Brazed seam	m	1.00	7.00	0.00	1.05	8.05

ZINC SHEET COVERINGS

	Unit	Labour hours	Net labour (£)	Net material (£)	O'heads /profit (£)	Total (£)
H74 ZINC SHEET COVERINGS						
12 gauge zinc (0.65mm thick) to BS849						
Roof covering						
flat	m2	3.00	21.00	10.34	4.70	36.04
sloping 10 to 50 degrees	m2	3.30	23.10	10.34	5.02	38.46
sloping or vertical over 50 degrees	m2	3.80	26.60	10.34	5.54	42.48
Flashing						
150mm girth	m	0.40	2.80	1.55	0.65	5.00
200mm girth	m	0.50	3.50	2.07	0.84	6.41
300mm girth	m	0.80	5.60	3.11	1.31	10.02
Stepped flashing						
150mm girth	m	0.50	3.50	1.55	0.76	5.81
200mm girth	m	0.60	4.20	2.07	0.94	7.21
300mm girth	m	0.90	6.30	3.11	1.41	10.82
Apron						
200mm girth	m	0.50	3.50	2.07	0.84	6.41
300mm girth	m	0.80	5.60	3.11	1.31	10.02
400mm girth	m	1.00	7.00	4.14	1.67	12.81
Capping to ridge or hip						
200mm girth	m	0.50	3.50	2.07	0.84	6.41
300mm girth	m	0.80	5.60	3.11	1.31	10.02
400mm girth	m	1.00	7.00	4.14	1.67	12.81
Lining to valley or gutter						
400mm girth	m	1.10	7.70	4.14	1.78	13.62
600mm girth	m	1.50	10.50	6.21	2.51	19.22
800mm girth	m	1.80	12.60	8.27	3.13	24.00
Raking cutting	m	0.10	0.70	0.00	0.10	0.80

RATES FOR MEASURED WORK

Zinc sheet coverings (cont'd)	Unit	Labour hours	Net labour (£)	Net material (£)	O'heads /profit (£)	Total (£)
Curved cutting	m	0.15	1.05	0.00	0.16	1.21
Welted edge	m	0.20	1.40	0.00	0.21	1.61
Beaded edge	m	0.20	1.40	0.00	0.21	1.61
Wedging into groove with zinc wedges	m	0.20	1.40	0.13	0.23	1.76
Soldered angle	m	1.00	7.00	0.00	1.05	8.05
Soldered seam	m	1.00	7.00	0.00	1.05	8.05

14 gauge zinc (0.8mm thick) to BS849

Roof covering

flat	m	3.00	21.00	12.19	4.98	38.17
sloping 10 to 50 degrees	m	3.30	23.10	12.19	5.29	40.58
sloping or vertical over 50 degrees	m	3.80	26.60	12.19	5.82	44.61

Flashing

150mm girth	m	0.40	2.80	1.83	0.69	5.32
200mm girth	m	0.50	3.50	2.44	0.89	6.83
300mm girth	m	0.80	5.60	3.65	1.39	10.64

Stepped flashing

150mm girth	m	0.50	3.50	1.83	0.80	6.13
200mm girth	m	0.60	4.20	2.44	1.00	7.64
300mm girth	m	0.90	6.30	3.65	1.49	11.44

Apron

200mm girth	m	0.50	3.50	2.44	0.89	6.83
300mm girth	m	0.80	5.60	3.65	1.39	10.64
400mm girth	m	1.00	7.00	4.87	1.78	13.65

Capping to ridge or hip

200mm girth	m	0.50	3.50	2.44	0.89	6.83

ZINC SHEET COVERINGS

	Unit	Labour hours	Net labour (£)	Net material (£)	O'heads /profit (£)	Total (£)
300mm girth	m	0.80	5.60	3.65	1.39	10.64
400mm girth	m	1.00	7.00	4.87	1.78	13.65
Lining to valley or gutter						
400mm girth	m	1.10	7.70	4.87	1.89	14.46
600mm girth	m	1.50	10.50	7.32	2.67	20.49
800mm girth	m	1.80	12.60	9.75	3.35	25.70
Raking cutting	m	0.10	0.70	0.00	0.10	0.80
Curved cutting	m	0.15	1.05	0.00	0.16	1.21
Welted edge	m	0.20	1.40	0.00	0.21	1.61
Beaded edge	m	0.20	1.40	0.00	0.21	1.61
Wedging into groove with zinc wedges	m	0.20	1.40	0.16	0.23	1.79
Soldered angle	m	1.00	7.00	0.00	1.05	8.05
Soldered seam	m	1.00	7.00	0.00	1.05	8.05

RATES FOR MEASURED WORK

	Unit	Labour hours	Net labour (£)	Net material (£)	O'heads /profit (£)	Total (£)
H76 FIBRE BITUMEN COVERINGS						
NURALITE						
Nuralite FX sheeting to roof coverings						
flat	m2	0.40	2.80	14.62	2.61	20.03
sloping 10 to 50 degrees	m2	0.45	3.15	14.62	2.67	20.44
vertical or sloping over 50 degrees	m2	0.85	5.95	14.62	3.09	23.66
gutters	m2	0.80	5.60	14.62	3.03	23.25
Beaded cover flashings, preformed, girth						
100mm	m	0.10	0.70	1.34	0.31	2.35
150mm	m	0.12	0.84	1.75	0.39	2.98
200mm	m	0.12	0.84	2.32	0.47	3.63
250mm	m	0.15	1.05	2.86	0.59	4.50
300mm	m	0.15	1.05	3.52	0.69	5.26
Ridge trays, preformed, girth						
250mm	nr	0.15	1.05	1.11	0.32	2.48
350mm	nr	0.15	1.05	1.31	0.35	2.71
450mm	nr	0.20	1.40	1.56	0.44	3.40
Intermediate trays, preformed, girth						
250mm	nr	0.15	1.05	1.29	0.35	2.69
350mm	nr	0.15	1.05	1.56	0.39	3.00
450mm	nr	0.20	1.40	1.71	0.47	3.58

FIBRE BITUMEN COVERINGS

	Unit	Labour hours	Net labour (£)	Net material (£)	O'heads /profit (£)	Total (£)
Catchment trays, preformed, girth						
250mm	nr	0.15	1.05	1.31	0.35	2.71
350mm	nr	0.15	1.05	1.71	0.41	3.17
450mm	nr	0.20	1.40	1.71	0.47	3.58
Cavity closure piece, girth						
250mm	nr	0.15	1.05	0.79	0.28	2.12
350mm	nr	0.15	1.05	0.96	0.30	2.31
450mm	nr	0.20	1.40	1.10	0.38	2.88
Linings to concrete gutters, preformed, girth						
450mm	m	0.20	1.40	6.03	1.11	8.54
490mm	m	0.20	1.40	6.03	1.11	8.54
500mm	m	0.20	1.40	6.03	1.11	8.54
519mm	m	0.20	1.40	6.03	1.11	8.54
Soakers, preformed, girth						
150mm	nr	0.05	0.35	0.59	0.14	1.08
175mm	nr	0.05	0.35	0.62	0.15	1.12
216mm	nr	0.10	0.70	0.76	0.22	1.68
250mm	nr	0.10	0.70	0.86	0.23	1.79
300mm	nr	0.15	1.05	1.00	0.31	2.36
350mm	nr	0.15	1.05	1.16	0.33	2.54
432mm	nr	0.20	1.40	1.74	0.47	3.61
Nuralite Nutec FX sheeting to roof coverings						
flat	m2	0.40	2.80	15.19	2.70	20.69
sloping 10 to 50 degrees	m2	0.45	3.15	15.19	2.75	21.09
vertical or sloping over 50 degrees	m2	0.85	5.95	15.19	3.17	24.31
gutters	m2	0.80	5.60	15.19	3.12	23.91

RATES FOR MEASURED WORK

Nuralite (cont'd)	Unit	Labour hours	Net labour (£)	Net material (£)	O'heads /profit (£)	Total (£)
Beaded cover flashings, preformed, girth						
100mm	m	0.10	0.70	1.34	0.31	2.35
150mm	m	0.12	0.84	1.75	0.39	2.98
200mm	m	0.12	0.84	2.32	0.47	3.63
250mm	m	0.15	1.05	2.86	0.59	4.50
300mm	m	0.15	1.05	3.52	0.69	5.26
Ridge trays, preformed, girth						
250mm	nr	0.15	1.05	1.11	0.32	2.48
350mm	nr	0.15	1.05	1.31	0.35	2.71
450mm	nr	0.20	1.40	1.56	0.44	3.40
Intermediate trays, preformed, girth						
250mm	nr	0.15	1.05	1.29	0.35	2.69
350mm	nr	0.15	1.05	1.56	0.39	3.00
450mm	nr	0.20	1.40	1.71	0.47	3.58
Catchment trays, preformed, girth						
250mm	nr	0.15	1.05	1.31	0.35	2.71
350mm	nr	0.15	1.05	1.71	0.41	3.17
450mm	nr	0.20	1.40	1.71	0.47	3.58
Cavity closure piece, girth						
250mm	nr	0.15	1.05	0.79	0.28	2.12
350mm	nr	0.15	1.05	0.96	0.30	2.31
450mm	nr	0.20	1.40	1.10	0.38	2.88
Linings to concrete gutters, preformed, girth						
450mm	m	0.20	1.40	6.03	1.11	8.54
490mm	m	0.20	1.40	6.03	1.11	8.54
500mm	m	0.20	1.40	6.03	1.11	8.54
519mm	m	0.20	1.40	6.03	1.11	8.54

FIBRE BITUMEN COVERINGS

	Unit	Labour hours	Net labour (£)	Net material (£)	O'heads /profit (£)	Total (£)
Soakers, preformed, girth						
150mm	nr	0.05	0.35	0.59	0.14	1.08
175mm	nr	0.05	0.35	0.62	0.15	1.12
216mm	nr	0.10	0.70	0.76	0.22	1.68
250mm	nr	0.10	0.70	0.86	0.23	1.79
300mm	nr	0.15	1.05	1.00	0.31	2.36
350mm	nr	0.15	1.05	1.16	0.33	2.54
432mm	nr	0.20	1.40	1.74	0.47	3.61

J WATERPROOFING

J41 Built up roofing

BUILT-UP ROOFING

	Unit	Labour hours	Net labour (£)	Net material (£)	O'heads /profit (£)	Total (£)

J41 BUILT-UP FELT ROOF COVERINGS

Built-up bituminous felt roof coverings, layers fully bonded with hot bitumen laid to 5 degrees pitch

Fibre based sand surfaced felt type 1B weighing 14kg/10m2

one layer	m2	0.22	1.54	1.45	0.45	3.44

Fibre based sand surfaced felt type 1B weighing 18kg/10m2

one layer	m2	0.23	1.61	1.72	0.50	3.83
two layers	m2	0.30	2.10	3.73	0.87	6.70
three layers	m2	0.45	3.15	5.73	1.33	10.21

Fibre based sand surfaced felt type 1B weighing 25kg/10m2

one layer	m2	0.24	1.68	2.28	0.59	4.55
two layers	m2	0.32	2.24	4.84	1.06	8.14
three layers	m2	0.47	3.29	7.51	1.62	12.42

Fibre based mineral surfaced felt type 1E weighing 38kg/10m2

one layer	m2	0.26	1.82	3.27	0.76	5.85

Glass fibre based sand surfaced felt type 3B weighing 18kg/10m2

one layer	m2	0.23	1.61	1.89	0.53	4.03
two layers	m2	0.30	2.10	4.06	0.92	7.08
three layers	m2	0.45	3.15	6.24	1.41	10.80

Glass fibre based mineral surfaced felt type 3E weighing 28kg/10m2

one layer	m2	0.26	1.82	2.60	0.66	5.08

RATES FOR MEASURED WORK

Felt roofing (cont'd)	Unit	Labour hours	Net labour (£)	Net material (£)	O'heads /profit (£)	Total (£)
Glass fibre based venting layer felt type 3G weighing 28kg/10m2						
one layer	m2	0.28	1.96	3.00	0.74	5.70
Polyester based sand surfaced felt type 5V weighing 29kg/10m2						
one layer	m2	0.25	1.75	6.80	1.28	9.83
Polyester based sand surfaced felt type 5B weighing 34kg/10m2						
one layer	m2	0.26	1.82	5.69	1.13	8.64
Polyester based mineral surfaced felt type 5E weighing 38kg/10m2						
one layer	m2	0.28	1.96	6.71	1.30	9.97
Polyester based (180g/m2) sand surfaced elastomeric bitumen coated felt weighing 40kg/20m2						
one layer	m2	0.24	1.68	3.76	0.82	6.26
Polyester based (180g/m2) mineral surfaced elastomeric bitumen coated felt weighing 32kg/10m2						
one layer	m2	0.28	1.96	5.03	1.05	8.04
Built-up bituminous felt roof coverings laid to 35 degrees pitch						
Fibre based and surfaced felt type 1B weighing 14kg/10m2						
one layer	m2	0.33	2.31	1.45	0.56	4.32

BUILT-UP ROOFING

	Unit	Labour hours	Net labour (£)	Net material (£)	O'heads /profit (£)	Total (£)
Fibre based sand surfaced felt type 1B weighing 18kg/10m2						
one layer	m2	0.35	2.45	1.72	0.63	4.80
two layers	m2	0.45	3.15	3.73	1.03	7.91
three layers	m2	0.68	4.76	5.73	1.57	12.06
Fibre based sand surfaced felt type 1B weighing 25kg/10m2						
one layer	m2	0.36	2.52	2.28	0.72	5.52
two layers	m2	0.48	3.36	4.84	1.23	9.43
three layers	m2	0.71	4.97	7.51	1.87	14.35
Fibre based mineral surfaced felt type 1E weighing 38kg/10m2						
one layer	m2	0.39	2.73	3.27	0.90	6.90
Glass fibre based sand surfaced felt type 3B weighing 18kg/10m2						
one layer	m2	0.35	2.45	1.89	0.65	4.99
two layers	m2	0.45	3.15	4.06	1.08	8.29
three layers	m2	0.68	4.76	6.26	1.65	12.67
Glass fibre based mineral surfaced felt type 3E weighing 28kg/10m2						
one layer	m2	0.39	2.73	2.60	0.80	6.13
Glass fibre based venting layer felt type 3G weighing 28kg/10m2						
one layer	m2	0.42	2.94	3.00	0.89	6.83
Polyester based sand surfaced felt type 5V weighing 29kg/10m2						
one layer	m2	0.35	2.45	6.80	1.39	10.64

RATES FOR MEASURED WORK

Felt roofing (cont'd)	Unit	Labour hours	Net labour (£)	Net material (£)	O'heads /profit (£)	Total (£)
Polyester based sand surfaced felt type 5B weighing 34kg/10m2					
one layer	m2	0.36	2.52	5.69	1.23	9.44
Polyester based mineral surfaced felt type 5E weighing 38kg/10m2					
one layer	m2	0.42	2.94	6.71	1.45	11.10
Polyester based (180g/m2) sand surfaced elastomeric bitumen coated felt weighing 40kg/20m2					
one layer	m2	0.34	2.38	3.76	0.92	7.06
Polyester based (180g/m2) mineral surfaced elastomeric bitumen coated felt weighing 32kg/10m2					
one layer	m2	0.36	2.52	5.03	1.13	8.68

Built-up bituminous felt roof coverings laid to 70 degrees pitch

Fibre based sand surfaced felt type 1B weighing 14kg/10m2 | | | | | |
---|---|---|---|---|---|---
one layer | m2 | 0.39 | 2.73 | 1.45 | 0.63 | 4.81
Fibre based sand surfaced felt type 1B weighing 18kg/10m2 | | | | | |
one layer | m2 | 0.40 | 2.80 | 1.72 | 0.68 | 5.20
two layers | m2 | 0.53 | 3.71 | 3.73 | 1.12 | 8.56
three layers | m2 | 0.79 | 5.53 | 5.73 | 1.69 | 12.95
Fibre based sand surfaced felt type 1B weighing 25kg/10m2 | | | | | |
one layer | m2 | 0.42 | 2.94 | 2.28 | 0.78 | 6.00
two layers | m2 | 0.56 | 3.92 | 4.84 | 1.31 | 10.07
three layers | m2 | 0.82 | 5.74 | 7.51 | 1.99 | 15.24

BUILT-UP ROOFING

	Unit	Labour hours	Net labour (£)	Net material (£)	O'heads /profit (£)	Total (£)
Fibre based mineral surfaced felt type 1E weighing 38kg/10m2						
one layer	m2	0.46	3.22	3.27	0.97	7.46
Glass fibre based sand surfaced felt type 3B weighing 18kg/10m2						
one layer	m2	0.40	2.80	1.89	0.70	5.39
two layers	m2	0.53	3.71	4.06	1.17	8.94
three layers	m2	0.79	5.53	6.24	1.77	13.54
Glass fibre based mineral surfaced felt type 3E weighing 28kg/10m2						
one layer	m2	0.46	3.22	2.60	0.87	6.69
Glass fibre based venting layer felt type 3G weighing 28kg/10m2						
one layer	m2	0.49	3.43	3.00	0.96	7.39
Polyester based sand surfaced felt type 5V weighing 29kg/10m2						
one layer	m2	0.40	2.80	6.80	1.44	11.04
Polyester based sand surfaced felt type 5B weighing 34kg/10m2						
one layer	m2	0.41	2.87	5.69	1.28	9.84
Polyester based mineral surfaced felt type 5E weighing 38kg/10m2						
one layer	m2	0.48	3.36	6.71	1.51	11.58
Polyester based (180g/m2) sand surfaced elatomeric bitumen coated felt weighing 40kg/10m2						
one layer	m2	0.39	2.73	3.76	0.97	7.46

RATES FOR MEASURED WORK

Felt roofing (cont'd)	Unit	Labour hours	Net labour (£)	Net material (£)	O'heads /profit (£)	Total (£)
Polyester based (180g/m2) mineral surfaced elastomeric bitumen coated felt weighing 32kg/10m2						
one layer	m2	0.41	2.87	5.03	1.19	9.09
Built-up bituminous felt roof coverings 0 to 200mm girth laid to 5 degrees pitch						
Fibre based sand surfaced felt type 1B weighing 14kg/10m2						
one layer	m	0.04	0.28	0.29	0.09	0.66
Fibre based sand surfaced felt type 1B weighing 18kg/10m2						
one layer	m	0.04	0.28	0.34	0.09	0.71
two layers	m	0.06	0.42	0.75	0.18	1.35
three layers	m	0.08	0.56	1.25	0.27	2.08
Fibre based sand surfaced felt type 1B weighing 25kg/10m2						
one layer	m	0.05	0.35	0.45	0.12	0.92
two layers	m	0.06	0.42	0.97	0.21	1.60
three layers	m	0.08	0.56	1.50	0.31	2.37
Fibre based mineral surfaced felt type 1E weighing 38kg/10m2						
one layer	m	0.05	0.35	0.65	0.15	1.15
Glass fibre based sand surfaced felt type 3B weighing 18kg/10m2						
one layer	m	0.04	0.28	0.38	0.10	0.76
two layers	m	0.06	0.42	0.81	0.18	1.41
three layers	m	0.08	0.56	1.25	0.27	2.08

BUILT-UP ROOFING

	Unit	Labour hours	Net labour (£)	Net material (£)	O'heads /profit (£)	Total (£)
Glass fibre based mineral surfaced felt type 3E weighing 28kg/10m2						
one layer	m	0.05	0.35	0.53	0.13	1.01
Glass fibre based venting layer felt type 3G weighing 28kg/10m2						
one layer	m	0.05	0.35	0.60	0.14	1.09
Polyester based sand surfaced felt type 5V weighing 29kg/10m2						
one layer	m	0.04	0.28	1.37	0.25	1.90
Polyester based sand surfaced felt type 5B weighing 34kg/10m2						
one layer	m	0.04	0.28	1.13	0.21	1.62
Polyester based mineral surfaced felt type 5E weighing 38kg/10m2						
one layer	m	0.05	0.35	1.34	0.25	1.94
Polyester based (180g/m2) sand surfaced elastomeric bitumen coated felt weighing 40kg/20m2						
one layer	m	0.06	0.42	0.76	0.18	1.36
Polyester based (180g/m2) mineral surfaced elastomeric bitumen coated felt weighing 32kg/10m2						
one layer	m	0.06	0.42	1.01	0.21	1.64

RATES FOR MEASURED WORK

Felt roofing (cont'd)	Unit	Labour hours	Net labour (£)	Net material (£)	O'heads /profit (£)	Total (£)
Built-up bituminous felt roof coverings 0 to 200mm girth laid to 35 degrees pitch						
Fibre based sand surfaced felt type 1B weighing 14kg/10m2						
one layer	m	0.06	0.42	0.29	0.11	0.82
Fibre based sand surfaced felt type 1B weighing 18kg/10m2						
one layer	m	0.07	0.49	0.34	0.12	0.95
two layers	m	0.08	0.56	0.75	0.20	1.51
three layers	m	0.13	0.91	1.14	0.31	2.36
Fibre based sand surfaced felt type 1B weighing 25kg/10m2						
one layer	m	0.07	0.49	0.45	0.14	1.08
two layers	m	0.09	0.63	0.97	0.24	1.84
three layers	m	0.13	0.91	1.50	0.36	2.77
Fibre based mineral surfaced felt type 1E weighing 38kg/10m2						
one layer	m	0.07	0.49	0.65	0.17	1.31
Glass fibre based sand surfaced felt type 3B weighing 18kg/10m2						
one layer	m	0.07	0.49	0.38	0.13	1.00
two layers	m	0.08	0.56	0.81	0.21	1.58
three layers	m	0.13	0.91	1.25	0.32	2.48
Glass fibre based mineral surfaced felt type 3E weighing 28kg/10m2						
one layer	m	0.07	0.49	0.53	0.15	1.17
Glass fibre based venting layer felt type 3G weighing 28kg/10m2						
one layer	m	0.08	0.56	0.60	0.17	1.33

BUILT-UP ROOFING

	Unit	Labour hours	Net labour (£)	Net material (£)	O'heads /profit (£)	Total (£)
Polyester based sand surfaced felt type 5V weighing 29kg/10m2						
one layer	m	0.06	0.42	1.37	0.27	2.06
Polyester based sand surfaced felt type 5B weighing 34kg/10m2						
one layer	m	0.06	0.42	1.13	0.23	1.78
Polyester based mineral surfaced felt type 5E weighing 38kg/10m2						
one layer	m	0.07	0.49	1.34	0.27	2.10
Polyester based (180g/m2) sand surfaced elastomeric bitumen costed felt weighing 40kg/10m2						
one layer	m	0.08	0.56	0.76	0.20	1.52
Polyester based (180g/m2) mineral surfaced elastomeric bitumen coated felt weighing 32kg/10m2						
one layer	m	0.08	0.56	1.01	0.24	1.81

Built-up bituminous felt roof coverings 0 to 200mm girth laid to 70 degrees pitch

	Unit	Labour hours	Net labour (£)	Net material (£)	O'heads /profit (£)	Total (£)
Fibre based sand surfaced felt type 1B weighing 14kg/10m2						
one layer	m	0.07	0.49	0.29	0.12	0.90
Fibre based sand surfaced felt type 1B weighing 18kg/10m2						
one layer	m	0.08	0.56	0.34	0.14	1.04
two layers	m	0.10	0.70	0.75	0.22	1.67
three layers	m	0.15	1.05	1.25	0.34	2.64

RATES FOR MEASURED WORK

Felt roofing (cont'd)	Unit	Labour hours	Net labour (£)	Net material (£)	O'heads /profit (£)	Total (£)
Fibre based sand surfaced felt type 1B weighing 25kg/10m2						
one layer	m	0.08	0.56	0.45	0.15	1.16
two layers	m	0.11	0.77	0.97	0.26	2.00
three layers	m	0.15	1.05	1.50	0.38	2.93
Fibre based mineral surfaced felt type 1E weighing 38kg/10m2						
one layer	m	0.09	0.63	0.65	0.19	1.47
Glass fibre based sand surfaced felt type 3B weighing 18kg/10m2						
one layer	m	0.08	0.56	0.38	0.14	1.08
two layers	m	0.10	0.70	0.81	0.23	1.74
three layers	m	0.15	1.05	1.25	0.34	2.64
Glass fibre based mineral surfaced felt type 3E weighing 28kg/10m2						
one layer	m	0.09	0.63	0.53	0.17	1.33
Glass fibre based venting layer felt type 3G weighing 26kg/10m2						
one layer	m	0.09	0.63	0.60	0.18	1.41
Polyester based sand surfaced felt type 5V weighing 29kg/10m2						
one layer	m	0.07	0.49	1.37	0.28	2.14
Polyester based sand surfaced felt type 5B weighing 34kg/10m2						
one layer	m	0.07	0.49	1.13	0.24	1.86
Polyester based mineral surfaced felt type 5E weighing 38kg/10m2						
one layer	m	0.08	0.56	1.34	0.29	2.19

BUILT-UP ROOFING

	Unit	Labour hours	Net labour (£)	Net material (£)	O'heads /profit (£)	Total (£)
Polyester based (180g/m2) sand surfaced elastomeric bitumen coated felt weighing 40kg/10m2						
one layer	m	0.09	0.63	0.76	0.21	1.60
Polyester based (180g/m2) mineral surfaced elastomeric bitumen coated felt weighing 32kg/10m2						
one layer	m	0.09	0.63	1.01	0.25	1.89
Built-up bituminous felt roof coverings 200 to 400mm girth laid to 5 degrees pitch						
Fibre based sand surfaced felt type 1B weighing 14kg/10m2						
one layer	m	0.08	0.56	0.58	0.17	1.31
Fibre based sand surfaced felt type 1B weighing 18kg/10m2						
one layer	m	0.08	0.56	0.68	0.19	1.43
two layers	m	0.10	0.70	1.49	0.33	2.52
three layers	m	0.16	1.12	2.29	0.51	3.92
Fibre based sand surfaced felt type 1B weighing 25kg/10m2						
one layer	m	0.08	0.56	0.91	0.22	1.69
two layers	m	0.11	0.77	1.93	0.40	3.10
three layers	m	0.16	1.12	3.00	0.62	4.74
Fibre based mineral surfaced felt type 1E weighing 38kg/10m2						
one layer	m	0.09	0.63	1.30	0.29	2.22

RATES FOR MEASURED WORK

Felt roofing (cont'd)	Unit	Labour hours	Net labour (£)	Net material (£)	O'heads /profit (£)	Total (£)
Glass fibre based sand surfaced felt type 3B weighing 18kg/10m2						
one layer	m	0.08	0.56	0.76	0.20	1.52
two layers	m	0.10	0.70	1.63	0.35	2.68
three layers	m	0.16	1.12	2.71	0.57	4.40
Glass fibre based mineral surfaced felt type 3E weighing 28kg/10m2						
one layer	m	0.09	0.63	1.04	0.25	1.92
Glass fibre based venting layer felt type 3G weighing 26kg/10m2						
one layer	m	0.10	0.70	1.20	0.28	2.18
Polyester based sand surfaced felt type 5V weighing 29kg/10m2						
one layer	m	0.08	0.56	2.72	0.49	3.77
Polyester based sand surfaced felt type 5B weighing 34kg/10m2						
one layer	m	0.08	0.56	2.28	0.43	3.27
Polyester based mineral surfaced felt type 5E weighing 38kg/10m2						
one layer	m	0.09	0.63	2.69	0.50	3.82
Polyester based (180g/m2) sand surfaced elastomeric bitumen coated felt weighing 40kg/10m2						
one layer	m	0.10	0.70	1.50	0.33	2.53
Polyester based (180g/m2) mineral surfaced elastomeric bitumen coated felt weighing 32kg/10m2						
one layer	m	0.10	0.70	2.02	0.41	3.13

BUILT-UP ROOFING

	Unit	Labour hours	Net labour (£)	Net material (£)	O'heads /profit (£)	Total (£)
Built-up bituminous felt roof coverings 200 to 400mm girth laid to 35 degrees pitch						
Fibre based sand surfaced felt 1B weighing 14kg/10m2						
one layer	m	0.11	0.77	0.58	0.20	1.55
Fibre based sand surfaced felt type 1B weighing 18kg/10m2						
one layer	m	0.12	0.84	0.68	0.23	1.75
two layers	m	0.16	1.12	1.49	0.39	3.00
three layers	m	0.23	1.61	2.29	0.58	4.48
Fibre based sand surfaced felt type 1B weighing 25kg/10m2						
one layer	m	0.12	0.84	0.91	0.26	2.01
two layers	m	0.17	1.19	1.93	0.47	3.59
three layers	m	0.24	1.68	3.00	0.70	5.38
Fibre based mineral surfaced felt type 1E weighing 38kg/10m2						
one layer	m	0.13	0.91	1.30	0.33	2.54
Glass fibre based sand surfaced felt type 3B weighing 18kg/10m2						
one layer	m	0.12	0.84	0.76	0.24	1.84
two layers	m	0.16	1.12	1.63	0.41	3.16
three layers	m	0.23	1.61	2.71	0.65	4.97
Glass fibre based mineral surfaced felt type 3E weighing 28kg/10m2						
one layer	m	0.13	0.91	1.04	0.29	2.24
Glass fibre based venting layer felt type 3G weighing 26kg/10m2						
one layer	m	0.14	0.98	1.51	0.37	2.86

RATES FOR MEASURED WORK

Felt roofing (cont'd)	Unit	Labour hours	Net labour (£)	Net material (£)	O'heads /profit (£)	Total (£)
Polyester based sand surfaced felt type 5V weighing 29kg/10m2						
one layer	m	0.12	0.84	2.72	0.53	4.09
Polyester based sand surfaced felt type 5B weighing 34kg/10m2						
one layer	m	0.12	0.84	2.28	0.47	3.59
Polyester based mineral surfaced felt type 5E weighing 38kg/10m2						
one layer	m	0.13	0.91	2.69	0.54	4.14
Polyester based (180g/m2) sand surfaced elastomeric bitumen coated felt weighing 40kg/10m2						
one layer	m	0.15	1.05	1.50	0.38	2.93
Polyester based (180g/m2) mineral surfaced elastomeric bitumen coated felt weighing 32kg/10m2						
one layer	m	0.15	1.05	2.02	0.46	3.53
Built-up bituminous felt roof coverings 200 to 400mm girth laid to 70 degrees pitch						
Fibre based sand surfaced felt type 1B weighing 14kg/10m2						
one layer	m	0.13	0.91	0.58	0.22	1.71
Fibre based sand surfaced felt type 1B weighing 18kg/10m2						
one layer	m	0.14	0.98	0.68	0.25	1.91
two layers	m	0.18	1.26	1.49	0.41	3.16
three layers	m	0.27	1.89	2.29	0.63	4.81

BUILT-UP ROOFING

	Unit	Labour hours	Net labour (£)	Net material (£)	O'heads /profit (£)	Total (£)
Fibre based sand surfaced felt type 1B weighing 25kg/10m2						
one layer	m	0.14	0.98	0.91	0.28	2.17
two layers	m	0.18	1.26	1.93	0.48	3.67
three layers	m	0.27	1.89	3.00	0.73	5.62
Fibre based mineral surfaced felt type 1E weighing 38kg/10m2						
one layer	m	0.16	1.12	1.30	0.36	2.78
Glass fibre based sand surfaced felt type 3B weighing 18kg/10m2						
one layer	m	0.14	0.98	0.76	0.26	2.00
two layers	m	0.18	1.26	1.63	0.43	3.32
three layers	m	0.27	1.89	2.71	0.69	5.29
Glass fibre based mineral surfaced felt type 3E weighing 28kg/10m2						
one layer	m	0.16	1.12	1.04	0.32	2.48
Glass fibre based venting layer felt type 3G weighing 26kg/10m2						
one layer	m	0.17	1.19	1.20	0.36	2.75
Polyester based sand surfaced felt type 5V weighing 29kg/10m2						
one layer	m	0.14	0.98	2.72	0.56	4.26
Polyester based sand surfaced felt type 5B weighing 34kg/10m2						
one layer	m	0.14	0.98	2.28	0.49	3.75
Polyester based mineral surfaced felt type 5E weighing 38kg/10m2						
one layer	m	0.15	1.05	2.69	0.56	4.30

RATES FOR MEASURED WORK

Felt roofing (cont'd)	Unit	Labour hours	Net labour (£)	Net material (£)	O'heads /profit (£)	Total (£)
Polyester based (180g/m2) sand surfaced elastomeric bitumen coted felt weighing 40kg/10m2						
one layer	m	0.17	1.19	1.50	0.40	3.09
Polyester based (180g/m2) mineral surfaced elastomeric bitumen coated felt weighing 32kg/10m2						
one layer	m	0.17	1.19	2.02	0.48	3.69
Working covering into outlets, gulleys or the like						
three-layer felt	nr	0.17	1.19	0.00	0.18	1.37
Labours						
raking cutting						
one layer	m	0.09	0.63	0.00	0.09	0.72
two layers	m	0.12	0.84	0.00	0.13	0.97
three layers	m	0.17	1.19	0.00	0.18	1.37
curved cutting						
one layer	m	0.11	0.77	0.00	0.12	0.89
two layers	m	0.14	0.98	0.00	0.15	1.13
three layers	m	0.20	1.40	0.00	0.21	1.61
Sundries						
6mm thick layer of limestone or granite chippings	m2	0.25	1.75	0.00	0.26	2.01
13mm thick layer of pea gravel	m2	0.32	2.24	0.00	0.34	2.58

BUILT-UP ROOFING

	Unit	Labour hours	Net labour (£)	Net material (£)	O'heads /profit (£)	Total (£)
Euroroof Ltd, high performance, elastomeric fully bonded roofing system						
Two-layer coverings, first layer HiTen G32, hot bitumen bonded to insulation layer, second layer, cap sheet HiTen sanded finish P45	m2	0.40	2.80	8.31	1.67	12.78
Top layer mineral surfaced	m2	0.10	0.70	1.01	0.26	1.97
10mm granite chippings	m2	0.10	0.70	3.79	0.67	5.16
Skirtings; two layers; top layer mineral surfaces; dressed over tilting fillet; turned into groove						
not exceeding 200mm girth	m	0.25	1.75	1.83	0.54	4.12
200 to 400mm girth	m	0.35	2.45	3.65	0.91	7.01
Linings to gutters; three layers						
400 to 600mm girth average	m	0.65	4.55	7.46	1.80	13.81
Collars around large pipes; two layers						
150mm high	nr	0.35	2.45	0.53	0.45	3.43
Eurovent breather	nr	0.30	2.10	9.27	1.71	13.08
Euroroof pipe seal flashing for						
pipes 50 to 75mm diameter	nr	0.30	2.10	14.54	2.50	19.14
pipes 100 to 150mm diameter	nr	0.40	2.80	16.75	2.93	22.48

L WINDOWS/DOORS/STAIRS

L11 Rooflights

ROOFLIGHTS

	Unit	Labour hours	Net labour (£)	Net material (£)	O'heads /profit (£)	Total (£)

L11 ROOFLIGHTS

'Coxdome Mark 1' rooflight plugged and screwed to builder's curb with 30 degrees sloped top as distributed by Coxdome Ltd

Single skin clear, diffused or tinted PVC-U domed rooflight

600mm diameter	nr	0.66	4.62	45.63	7.54	57.79
900mm diameter	nr	0.66	4.62	66.08	10.60	81.30
1200mm diameter	nr	0.83	5.81	83.46	13.39	102.66
1800mm diameter	nr	0.83	5.81	213.63	32.92	252.36

Single skin clear or diffused wire laminate PVC-U domed rooflight

600mm diameter	nr	0.66	4.62	55.21	8.97	68.80
900mm diameter	nr	0.66	4.62	79.32	12.59	96.53
1200mm diameter	nr	0.83	5.81	100.84	16.00	122.65

Single skin clear or diffused polycarbonate domed rooflight

600mm diameter	nr	0.66	4.62	64.68	10.40	79.70
900mm diameter	nr	0.66	4.62	91.64	14.44	110.70
1200mm diameter	nr	0.83	5.81	118.23	18.61	142.65
1800mm diameter	nr	0.83	5.81	303.02	46.32	355.15

Note: Add 20% to basic material costs for single skin tinted acrylic domed rooflight

Double skin domed rooflight with clear PVC-U standard inner skin and clear, diffused or tinted PVC-U outer skin

600mm diameter	nr	0.66	4.62	80.56	12.78	97.96
900mm diameter	nr	0.66	4.62	120.97	18.84	144.43
1200mm diameter	nr	0.83	5.81	157.08	24.43	187.32
1800mm diameter	nr	0.83	5.81	402.24	61.21	469.26

RATES FOR MEASURED WORK

'Coxdome Mark 1' (cont'd)	Unit	Labour hours	Net labour (£)	Net material (£)	O'heads /profit (£)	Total (£)
Double skin domed rooflight with clear PVC-U standard inner skin and clear or diffused wire laminate PVC-U outer skin						
600mm diameter	nr	0.66	4.62	97.29	15.29	117.20
900mm diameter	nr	0.66	4.62	145.18	22.47	172.27
1200mm diameter	nr	0.83	5.81	189.80	29.34	224.95
Double skin domed rooflight with clear PVC-U standard inner skin and clear or diffused polycarbonate outer skin						
600mm diameter	nr	0.66	4.62	105.68	16.55	126.85
900mm diameter	nr	0.66	4.62	169.41	26.10	200.13
1200mm diameter	nr	0.83	5.81	222.52	34.25	262.58
1800mm diameter	nr	0.83	5.81	569.88	86.35	662.04
Note: Add 20% to basic material costs for tinted acrylic outer skin						
'Coxdome Mark 2' rooflight plugged and screwed to flat builder's curb as distributed by Coxdome Ltd						
Single skin clear, diffused or tinted PVC-U domed rooflight						
619 x 619mm	nr	1.00	7.00	46.98	8.10	62.08
772 x 772mm	nr	1.08	7.56	54.83	9.36	71.75
924 x 924mm	nr	1.16	8.12	62.75	10.63	81.50
1076 x 1076mm	nr	1.45	10.15	62.91	10.96	84.02
1229 x 1229mm	nr	1.66	11.62	151.91	24.53	188.06
Single skin clear, diffused or tinted PVC-U pyramidal rooflight						
619 x 619mm	nr	1.00	7.00	84.11	13.67	104.78
772 x 772mm	nr	1.08	7.56	91.59	14.87	114.02
924 x 924mm	nr	1.33	9.31	99.02	16.25	124.58
1076 x 1076mm	nr	1.45	10.15	101.27	16.71	128.13
1229 x 1229mm	nr	1.66	11.62	186.84	29.77	228.23

ROOFLIGHTS

	Unit	Labour hours	Net labour (£)	Net material (£)	O'heads /profit (£)	Total (£)
Single skin clear or diffused wire laminate PVC-U domed rooflight						
619 x 619mm	nr	1.00	7.00	68.12	11.27	86.39
772 x 772mm	nr	1.16	8.12	79.54	13.15	100.81
924 x 924mm	nr	1.33	9.31	90.99	15.04	115.34
1076 x 1076mm	nr	1.45	10.15	91.27	15.21	116.63
Single skin clear or diffused polycarbonate domed rooflight						
619 x 619mm	nr	1.00	7.00	79.92	13.04	99.96
772 x 772mm	nr	1.16	8.12	93.04	15.17	116.33
924 x 924mm	nr	1.33	9.31	106.71	17.40	133.42
1076 x 1076mm	nr	1.45	10.15	106.98	17.57	134.70
1229 x 1229mm	nr	1.66	11.62	258.30	40.49	310.41
Single skin clear or diffused polycarbonate pyramidal rooflight						
619 x 619mm	nr	1.00	7.00	143.14	22.52	172.66
772 x 772mm	nr	1.16	8.12	154.71	24.42	187.25
924 x 924mm	nr	1.33	9.31	168.33	26.65	204.29
1076 x 1076mm	nr	1.45	10.15	172.20	27.35	209.70
1229 x 1229mm	nr	1.66	11.62	317.50	49.37	378.49

Note: Add 20% to basic material costs for single skin tinted acrylic domed or pyramidal rooflight

Double skin domed rooflight with clear PVC-U standard inner skin and clear, diffused or tinted acrylic outer skin						
619 x 619mm	nr	1.00	7.00	82.66	13.45	103.11
772 x 772mm	nr	1.16	8.12	98.21	15.95	122.28
924 x 924mm	nr	1.33	9.31	113.81	18.47	141.59
1076 x 1076mm	nr	1.45	10.15	115.37	18.83	144.35
1229 x 1229mm	nr	1.66	11.62	312.55	48.63	372.80

RATES FOR MEASURED WORK

'Coxdome Mark 2' (cont'd)	Unit	Labour hours	Net labour (£)	Net material (£)	O'heads /profit (£)	Total (£)
Double skin pyramidal rooflight with PVC-U standard inner skin and clear, diffused or tinted PVC-U outer skin						
619 x 619mm	nr	1.00	7.00	123.50	19.57	150.07
772 x 772mm	nr	1.16	8.12	138.73	22.03	168.88
924 x 924mm	nr	1.33	9.31	154.01	24.50	187.82
1076 x 1076mm	nr	1.45	10.15	159.34	25.42	194.91
1229 x 1229mm	nr	1.66	11.62	330.52	51.32	393.46
Double skin domed rooflight with clear PVC-U standard inner skin and clear or diffused wire laminate PVC-U outer skin						
619 x 619mm	nr	1.00	7.00	119.90	19.04	145.94
772 x 772mm	nr	1.16	8.12	142.44	22.58	173.14
924 x 924mm	nr	1.33	9.31	165.04	26.15	200.50
1076 x 1076mm	nr	1.45	10.15	167.31	26.62	204.08
Double skin domed rooflight with clear PVC-U standard inner skin and clear or diffused polycarbonate outer skin						
619 x 619mm	nr	1.00	7.00	140.50	22.12	169.62
772 x 772mm	nr	1.16	8.12	166.39	26.18	200.69
924 x 924mm	nr	1.33	9.31	193.52	30.42	233.25
1076 x 1076mm	nr	1.45	10.15	196.15	30.95	237.25
1229 x 1229mm	nr	1.66	11.62	499.22	76.63	587.47
Double skin pyramidal rooflight with clear PVC-U standard inner skin and clear or diffused acrylic outer skin						
619 x 619mm	nr	1.00	7.00	148.20	23.28	178.48
772 x 772mm	nr	1.16	8.12	165.74	26.08	199.94
924 x 924mm	nr	1.33	9.31	184.84	29.12	223.27
1076 x 1076mm	nr	1.45	10.15	191.19	30.20	231.54
1229 x 1229mm	nr	1.66	11.62	396.60	61.23	469.45

ROOFLIGHTS

	Unit	Labour hours	Net labour (£)	Net material (£)	O'heads /profit (£)	Total (£)

Note: Add 20% to basic material costs for tinted acrylic outer skin

'Coxdome Mark 3' rooflight plugged and screwed to builder's curb with 30 degrees sloped top as distributed by Coxdome Ltd

Single skin clear, diffused or tinted acrylic domed rooflight

600 x 600mm	nr	1.00	7.00	36.87	6.58	50.45
900 x 600mm	nr	1.16	8.12	64.35	10.87	83.34
900 x 900mm	nr	1.33	9.31	71.25	12.08	92.64
1200 x 900mm	nr	1.50	10.50	75.61	12.92	99.03
1200 x 1200mm	nr	1.66	11.62	105.47	17.56	134.65
1800 x 1200mm	nr	2.00	14.00	174.08	28.21	216.29

Single skin clear, diffused or tinted acrylic pyramidal rooflight

600 x 600mm	nr	1.00	7.00	40.57	7.14	54.71
900 x 600mm	nr	1.16	8.12	70.76	11.83	90.71
900 x 900mm	nr	1.33	9.31	78.35	13.15	100.81
1200 x 900mm	nr	1.50	10.50	92.34	15.43	118.27
1200 x 1200mm	nr	1.66	11.62	116.01	19.14	146.77
1800 x 1200mm	nr	2.00	14.00	191.52	30.83	236.35

Single skin clear or diffused wire laminate PVC-U domed rooflight

600 x 600mm	nr	1.00	7.00	43.05	7.51	57.56
900 x 600mm	nr	1.16	8.12	89.33	14.62	112.07
900 x 900mm	nr	1.33	9.31	98.37	16.15	123.83
1200 x 900mm	nr	1.50	10.50	110.16	18.10	138.76
1200 x 1200mm	nr	1.66	11.62	130.44	21.31	163.37
1800 x 1200mm	nr	2.00	14.00	193.94	31.19	239.13

RATES FOR MEASURED WORK

'Coxdome Mark 3' (cont'd)	Unit	Labour hours	Net labour (£)	Net material (£)	O'heads /profit (£)	Total (£)
Single skin clear or diffused PVC-U domed rooflight						
600 x 600mm	nr	1.00	7.00	34.60	6.24	47.84
900 x 600mm	nr	1.16	8.12	71.46	11.94	91.52
900 x 900mm	nr	1.33	9.31	71.89	12.18	93.38
1200 x 900mm	nr	1.50	10.50	86.96	14.62	112.08
1200 x 1200mm	nr	1.66	11.62	90.25	15.28	117.15
1800 x 1200mm	nr	2.00	14.00	153.04	25.06	192.10
Single skin clear or diffused PVC-U pyramidal rooflight						
600 x 600mm	nr	1.00	7.00	38.04	6.76	51.80
900 x 600mm	nr	1.16	8.12	78.68	13.02	99.82
900 x 900mm	nr	1.33	9.31	79.05	13.25	101.61
1200 x 900mm	nr	1.50	10.50	95.73	15.93	122.16
1200 x 1200mm	nr	1.66	11.62	99.39	16.65	127.66
1800 x 1200mm	nr	2.00	14.00	168.33	27.35	209.68

Note: Add 20% to basic material costs for single skin tinted acrylic domed or pyramidal rooflight

	Unit	Labour hours	Net labour (£)	Net material (£)	O'heads /profit (£)	Total (£)
Single skin polycarbonate domed rooflight						
600 x 600mm	nr	1.00	7.00	62.27	10.39	79.66
900 x 600mm	nr	1.16	8.12	128.72	20.53	157.37
900 x 900mm	nr	1.33	9.31	129.42	20.81	159.54
1200 x 900mm	nr	1.50	10.50	156.43	25.04	191.97
1200 x 1200mm	nr	1.66	11.62	162.46	26.11	200.19
1800 x 1200mm	nr	2.00	14.00	275.47	43.42	332.89
Single skin polycarbonate pyramidal rooflight						
600 x 600mm	nr	1.00	7.00	68.50	11.32	86.82
900 x 600mm	nr	1.16	8.12	206.16	32.14	246.42
900 x 900mm	nr	1.33	9.31	142.28	22.74	174.33
1200 x 900mm	nr	1.50	10.50	172.25	27.41	210.16
1200 x 1200mm	nr	1.66	11.62	178.66	28.54	218.82
1800 x 1200mm	nr	2.00	14.00	303.02	47.55	364.57

ROOFLIGHTS

	Unit	Labour hours	Net labour (£)	Net material (£)	O'heads /profit (£)	Total (£)

Double skin domed rooflight with clear PVC-U standard inner skin and clear, diffused or tinted acrylic outer skin

600 x 600mm	nr	1.00	7.00	75.44	12.37	94.81
900 x 600mm	nr	1.16	8.12	111.66	17.97	137.75
900 x 900mm	nr	1.33	9.31	140.67	22.50	172.48
1200 x 900mm	nr	1.50	10.50	161.68	25.83	198.01
1200 x 1200mm	nr	1.66	11.62	196.15	31.17	238.94
1800 x 1200mm	nr	2.00	14.00	299.41	47.01	360.42

Double skin pyramidal rooflight with clear PVC-U standard inner skin and clear, diffused or tinted acrylic outer skin

600 x 600mm	nr	1.00	7.00	86.74	14.06	107.80
900 x 600mm	nr	1.16	8.12	128.45	20.49	157.06
900 x 900mm	nr	1.33	9.31	161.76	25.66	196.73
1200 x 900mm	nr	1.50	10.50	184.68	29.28	224.46
1200 x 1200mm	nr	1.66	11.62	225.53	35.57	272.72
1800 x 1200mm	nr	2.00	14.00	333.58	52.14	399.72

Double skin domed rooflight with clear PVC-U standard inner skin and clear or diffused wire laminate PVC-U outer skin

600 x 600mm	nr	1.00	7.00	87.39	14.16	108.55
900 x 600mm	nr	1.16	8.12	152.50	24.09	184.71
900 x 900mm	nr	1.33	9.31	184.79	29.11	223.21
1200 x 900mm	nr	1.50	10.50	185.70	29.43	225.63
1200 x 1200mm	nr	1.66	11.62	236.40	37.20	285.22
1800 x 1200mm	nr	2.00	14.00	341.11	53.27	408.38

RATES FOR MEASURED WORK

'Coxdome Mark 3' (cont'd)	Unit	Labour hours	Net labour (£)	Net material (£)	O'heads /profit (£)	Total (£)

Double skin domed rooflight with clear PVC-U standard inner skin and clear or diffused PVC-U outer skin

600 x 600mm	nr	1.00	7.00	59.68	10.00	76.68
900 x 600mm	nr	1.16	8.12	117.42	18.83	144.37
900 x 900mm	nr	1.33	9.31	119.09	19.26	147.66
1200 x 900mm	nr	1.50	10.50	133.83	21.65	165.98
1200 x 1200mm	nr	1.66	11.62	145.18	23.52	180.32
1800 x 1200mm	nr	2.00	14.00	271.70	42.85	328.55

Double skin pyramidal rooflight with clear PVC-U standard inner skin and clear or diffused PVC-U outer skin

600 x 600mm	nr	1.00	7.00	68.55	11.33	86.88
900 x 600mm	nr	1.16	8.12	135.07	21.48	164.67
900 x 900mm	nr	1.33	9.31	136.90	21.93	168.14
1200 x 900mm	nr	1.50	10.50	153.90	24.66	189.06
1200 x 1200mm	nr	1.66	11.62	166.98	26.79	205.39
1800 x 1200mm	nr	2.00	14.00	312.44	48.97	375.41

Note: Add 20% to basic material costs for tinted acrylic outer skin

Double skin domed rooflight with clear PVC-U standard inner skin and clear, diffused or tinted polycarbonate outer skin

600 x 600mm	nr	1.00	7.00	107.47	17.17	131.64
900 x 600mm	nr	1.16	8.12	174.63	27.41	210.16
900 x 900mm	nr	1.33	9.31	214.34	33.55	257.20
1200 x 900mm	nr	1.50	10.50	228.38	35.83	274.71
1200 x 1200mm	nr	1.66	11.62	290.75	45.36	347.73
1800 x 1200mm	nr	2.00	14.00	387.72	60.26	461.98

ROOFLIGHTS

	Unit	Labour hours	Net labour (£)	Net material (£)	O'heads /profit (£)	Total (£)

Double skin pyramidal rooflight with clear PVC-U standard inner skin and clear, diffused or tinted polycarbonate outer skin

600 x 600mm	nr	1.00	7.00	123.55	19.58	150.13
900 x 600mm	nr	1.16	8.12	200.77	31.33	240.22
900 x 900mm	nr	1.33	9.31	246.46	38.37	294.14
1200 x 900mm	nr	1.50	10.50	263.26	41.06	314.82
1200 x 1200mm	nr	1.66	11.62	334.34	51.89	397.85
1800 x 1200mm	nr	2.00	14.00	453.26	70.09	537.35

'Coxdome Mark 4' rooflight fixed to insulated GRP upstand screwed to roof deck as distributed by Coxdome Ltd

Single skin clear, diffused or tinted PVC-U domed rooflight

600 x 600mm	nr	1.00	7.00	92.94	14.99	114.93
900 x 600mm	nr	1.16	8.12	157.72	24.88	190.72
900 x 900mm	nr	1.33	9.31	166.17	26.32	201.80
1200 x 900mm	nr	1.50	10.50	204.70	32.28	247.48
1200 x 1200mm	nr	1.66	11.62	217.73	34.40	263.75
1800 x 1200mm	nr	2.00	14.00	325.13	50.87	390.00

Single skin clear, diffused or tinted PVC-U pyramidal rooflight

600 x 600mm	nr	1.00	7.00	96.38	15.51	118.89
900 x 600mm	nr	1.16	8.12	164.93	25.96	199.01
900 x 900mm	nr	1.33	9.31	173.33	27.40	210.04
1200 x 900mm	nr	1.50	10.50	213.48	33.60	257.58
1200 x 1200mm	nr	1.66	11.62	226.77	35.76	274.15
1800 x 1200mm	nr	2.00	14.00	340.42	53.16	407.58

RATES FOR MEASURED WORK

'Coxdome Mark 4' (cont'd)	Unit	Labour hours	Net labour (£)	Net material (£)	O'heads /profit (£)	Total (£)
Single skin clear or diffused wire laminate PVC-U domed rooflight						
600 x 600mm	nr	1.00	7.00	101.39	16.26	124.65
900 x 600mm	nr	1.16	8.12	175.59	27.56	211.27
900 x 900mm	nr	1.33	9.31	192.64	30.29	232.24
1200 x 900mm	nr	1.50	10.50	227.77	35.74	274.01
1200 x 1200mm	nr	1.66	11.62	248.88	39.07	299.57
1800 x 1200mm	nr	2.00	14.00	380.61	59.19	453.80
Single skin clear or diffused acrylic domed rooflight						
600 x 600mm	nr	1.00	7.00	94.12	15.17	116.29
900 x 600mm	nr	1.16	8.12	150.62	23.81	182.55
900 x 900mm	nr	1.33	9.31	165.53	26.23	201.07
1200 x 900mm	nr	1.50	10.50	201.53	31.80	243.83
1200 x 1200mm	nr	1.66	11.62	223.91	35.33	270.86
1800 x 1200mm	nr	2.00	14.00	360.76	56.21	430.97
Single skin clear or diffused acrylic pyramidal rooflight						
600 x 600mm	nr	1.00	7.00	97.94	15.74	120.68
900 x 600mm	nr	1.16	8.12	157.03	24.77	189.92
900 x 900mm	nr	1.33	9.31	172.63	27.29	209.23
1200 x 900mm	nr	1.50	10.50	209.98	33.07	253.55
1200 x 1200mm	nr	1.66	11.62	234.47	36.91	283.00
1800 x 1200mm	nr	2.00	14.00	378.20	58.83	451.03
Note: Add 20% to basic material costs for single skin tinted acrylic domed or pyramidal rooflight						
Double skin domed rooflight with clear PVC-U standard inner skin and clear, diffused or tinted PVC-U outer skin						
600 x 600mm	nr	1.00	7.00	107.25	17.14	131.39
900 x 600mm	nr	1.16	8.12	203.68	31.77	243.57
900 x 900mm	nr	1.33	9.31	213.37	33.40	256.08
1200 x 900mm	nr	1.50	10.50	250.02	39.08	299.60

ROOFLIGHTS

	Unit	Labour hours	Net labour (£)	Net material (£)	O'heads /profit (£)	Total (£)
1200 x 1200mm	nr	1.66	11.62	263.62	41.29	316.53
1800 x 1200mm	nr	2.00	14.00	458.48	70.87	543.35

Double skin pyramidal rooflight with clear PVC-U standard inner skin and clear, diffused or tinted PVC-U outer skin

600 x 600mm	nr	1.00	7.00	116.13	18.47	141.60
900 x 600mm	nr	1.16	8.12	221.33	34.42	263.87
900 x 900mm	nr	1.33	9.31	231.18	36.07	276.56
1200 x 900mm	nr	1.50	10.50	270.30	42.12	322.92
1200 x 1200mm	nr	1.66	11.62	285.42	44.56	341.60
1800 x 1200mm	nr	2.00	14.00	499.11	76.97	590.08

Double skin domed rooflight with clear PVC-U standard inner skin and clear or diffused wire laminate PVC-U outer skin

600 x 600mm	nr	1.00	7.00	145.72	22.91	175.63
900 x 600mm	nr	1.16	8.12	238.77	37.03	283.92
900 x 900mm	nr	1.33	9.31	279.07	43.26	331.64
1200 x 900mm	nr	1.50	10.50	303.34	47.08	360.92
1200 x 1200mm	nr	1.66	11.62	354.84	54.97	421.43
1800 x 1200mm	nr	2.00	14.00	513.21	79.08	606.29

Double skin domed rooflight with clear PVC-U standard inner skin and clear or diffused acrylic outer skin

600 x 600mm	nr	1.00	7.00	133.78	21.12	161.90
900 x 600mm	nr	1.16	8.12	197.93	30.91	236.96
900 x 900mm	nr	1.33	9.31	234.95	36.64	280.90
1200 x 900mm	nr	1.50	10.50	278.26	43.31	332.07
1200 x 1200mm	nr	1.66	11.62	314.59	48.93	375.14
1800 x 1200mm	nr	2.00	14.00	486.09	75.01	575.10

RATES FOR MEASURED WORK

'Coxdome Mark 4' (cont'd)	Unit	Labour hours	Net labour (£)	Net material (£)	O'heads /profit (£)	Total (£)

Double skin pyramidal rooflight with clear PVC-U standard inner skin and clear or diffused acrylic outer skin

600 x 600mm	nr	1.00	7.00	145.08	22.81	174.89
900 x 600mm	nr	1.16	8.12	214.71	33.42	256.25
900 x 900mm	nr	1.33	9.31	256.04	39.80	305.15
1200 x 900mm	nr	1.50	10.50	302.32	46.92	359.74
1200 x 1200mm	nr	1.66	11.62	343.97	53.34	408.93
1800 x 1200mm	nr	2.00	14.00	531.02	81.75	626.77

Note: Add 20% to basic material costs for tinted acrylic outer skin

'Coxdome Mark 4' rooflight fixed to aluminium 'hit and miss' ventilator with black PVC shutters screwed to builder's curb with 30 degree sloped top as distributed by Coxdome Ltd

Single skin clear, diffused or tinted PVC-U domed rooflight

600 x 600mm	nr	1.00	7.00	153.10	24.01	184.11
900 x 600mm	nr	1.16	8.12	212.82	33.14	254.08
900 x 900mm	nr	1.33	9.31	230.75	36.01	276.07
1200 x 900mm	nr	1.50	10.50	277.35	43.18	331.03
1200 x 1200mm	nr	1.66	11.62	198.46	31.51	241.59
1800 x 1200mm	nr	2.00	14.00	420.92	65.24	500.16

Single skin clear, diffused or tinted PVC-U pyramidal rooflight

600 x 600mm	nr	1.00	7.00	156.54	24.53	188.07
900 x 600mm	nr	1.16	8.12	220.04	34.22	262.38
900 x 900mm	nr	1.33	9.31	238.01	37.10	284.42
1200 x 900mm	nr	1.50	10.50	286.10	44.49	341.09
1200 x 1200mm	nr	1.66	11.62	207.07	32.80	251.49
1800 x 1200mm	nr	2.00	14.00	436.20	67.53	517.73

ROOFLIGHTS

	Unit	Labour hours	Net labour (£)	Net material (£)	O'heads /profit (£)	Total (£)

Single skin clear or diffused wire laminate PVC-U domed rooflight

600 x 600mm	nr	1.00	7.00	161.54	25.28	193.82
900 x 600mm	nr	1.16	8.12	230.70	35.82	274.64
900 x 900mm	nr	1.33	9.31	257.22	39.98	306.51
1200 x 900mm	nr	1.50	10.50	321.00	49.73	381.23
1200 x 1200mm	nr	1.66	11.62	356.18	55.17	422.97
1800 x 1200mm	nr	2.00	14.00	461.60	71.34	546.94

Single skin clear or diffused acrylic domed rooflight

600 x 600mm	nr	1.00	7.00	155.36	24.35	186.71
900 x 600mm	nr	1.16	8.12	205.73	32.08	245.93
900 x 900mm	nr	1.33	9.31	230.11	35.91	275.33
1200 x 900mm	nr	1.50	10.50	293.98	45.67	350.15
1200 x 1200mm	nr	1.66	11.62	218.54	34.52	264.68
1800 x 1200mm	nr	2.00	14.00	441.75	68.36	524.11

Single skin clear or diffused acrylic pyramidal rooflight

600 x 600mm	nr	1.00	7.00	159.18	24.93	191.11
900 x 600mm	nr	1.16	8.12	212.13	33.04	253.29
900 x 900mm	nr	1.33	9.31	237.21	36.98	283.50
1200 x 900mm	nr	0.00	0.00	302.32	45.35	347.67
1200 x 1200mm	nr	1.66	11.62	229.08	36.10	276.80
1800 x 1200mm	nr	2.00	14.00	459.19	70.98	544.17

Note: Add 20% to basic material costs for single skin tinted acrylic domed or pyramidal rooflight

Double skin domed rooflight with clear PVC-U standard inner skin and clear, diffused or tinted PVC-U outer skin

600 x 600mm	nr	1.00	7.00	178.17	27.78	212.95
900 x 600mm	nr	1.16	8.12	258.78	40.03	306.93
900 x 900mm	nr	1.33	9.31	277.95	43.09	330.35
1200 x 900mm	nr	1.50	10.50	324.22	50.21	384.93
1200 x 1200mm	nr	1.66	11.62	373.35	57.75	442.72

RATES FOR MEASURED WORK

'Coxdome Mark 4' (cont'd)	Unit	Labour hours	Net labour (£)	Net material (£)	O'heads /profit (£)	Total (£)
1800 x 1200mm	nr	2.00	14.00	580.10	89.11	683.21

Double skin pyramidal rooflight with clear PVC-U standard inner skin and clear, diffused or tinted PVC-U outer skin

600 x 600mm	nr	1.00	7.00	187.06	29.11	223.17
900 x 600mm	nr	1.16	8.12	276.43	42.68	327.23
900 x 900mm	nr	1.33	9.31	295.86	45.78	350.95
1200 x 900mm	nr	1.50	10.50	344.19	53.20	407.89
1200 x 1200mm	nr	1.66	11.62	395.15	61.02	467.79
1800 x 1200mm	nr	2.00	14.00	610.07	93.61	717.68

Double skin domed rooflight with clear PVC-U standard inner skin and clear or diffused wire laminate PVC-U outer skin

600 x 600mm	nr	1.00	7.00	205.88	31.93	244.81
900 x 600mm	nr	1.16	8.12	293.87	45.30	347.29
900 x 900mm	nr	1.33	9.31	343.64	52.94	405.89
1200 x 900mm	nr	1.50	10.50	376.14	58.00	444.64
1200 x 1200mm	nr	1.66	11.62	442.77	68.16	522.55
1800 x 1200mm	nr	2.00	14.00	651.83	99.87	765.70

Double skin domed rooflight with clear PVC-U standard inner skin and clear or diffused acrylic outer skin

600 x 600mm	nr	1.00	7.00	193.94	30.14	231.08
900 x 600mm	nr	1.16	8.12	253.03	39.17	300.32
900 x 900mm	nr	1.33	9.31	299.52	46.32	355.15
1200 x 900mm	nr	1.50	10.50	351.02	54.23	415.75
1200 x 1200mm	nr	1.66	11.62	402.52	62.12	476.26
1800 x 1200mm	nr	2.00	14.00	549.85	84.58	648.43

ROOFLIGHTS

	Unit	Labour hours	Net labour (£)	Net material (£)	O'heads /profit (£)	Total (£)

Double skin pyramidal rooflight with clear PVC-U standard inner skin and clear or diffused acrylic outer skin

600 x 600mm	nr	1.00	7.00	205.24	31.84	244.08
900 x 600mm	nr	1.16	8.12	269.82	41.69	319.63
900 x 900mm	nr	1.33	9.31	319.97	49.39	378.67
1200 x 900mm	nr	1.50	10.50	374.97	57.82	443.29
1200 x 1200mm	nr	1.66	11.62	431.90	66.53	510.05
1800 x 1200mm	nr	2.00	14.00	594.79	91.32	700.11

Note: Add 20% to basic material costs for tinted acrylic outer skin

'Coxdome Mark 4' rooflight fixed to Slimline (C/050) ventilator unit comprising top hung dampers within extruded aluminium sections including aluminium splayed upstand screwed to roof deck as distributed by Coxdome Ltd

Single skin clear, diffused or tinted PVC-U domed rooflight

600 x 600mm	nr	1.00	7.00	178.50	27.82	213.32
900 x 600mm	nr	1.16	8.12	233.28	36.21	277.61
900 x 900mm	nr	1.33	9.31	275.61	42.74	327.66
1200 x 900mm	nr	1.50	10.50	296.94	46.12	353.56
1200 x 1200mm	nr	1.66	11.62	310.93	48.38	370.93
1800 x 1200mm	nr	2.00	14.00	419.79	65.07	498.86

Single skin clear, diffused or tinted PVCU pyramidal rooflight

600 x 600mm	nr	1.00	7.00	181.94	28.34	217.28
900 x 600mm	nr	1.16	8.12	240.38	37.27	285.77
900 x 900mm	nr	1.33	9.31	283.92	43.98	337.21
1200 x 900mm	nr	1.50	10.50	305.71	47.43	363.64
1200 x 1200mm	nr	1.66	11.62	319.97	49.74	381.33
1800 x 1200mm	nr	2.00	14.00	435.08	67.36	516.44

RATES FOR MEASURED WORK

'Coxdome Mark 4' (cont'd)	Unit	Labour hours	Net labour (£)	Net material (£)	O'heads /profit (£)	Total (£)
Single skin clear or diffused wire laminate PVC-U domed rooflight						
600 x 600mm	nr	1.00	7.00	191.25	29.74	227.99
900 x 600mm	nr	1.16	8.12	251.14	38.89	298.15
900 x 900mm	nr	1.33	9.31	291.29	45.09	345.69
1200 x 900mm	nr	1.50	10.50	320.60	49.66	380.76
1200 x 1200mm	nr	1.66	11.62	351.13	54.41	417.16
1800 x 1200mm	nr	2.00	14.00	460.69	71.20	545.89
Single skin clear or diffused acrylic domed rooflight						
600 x 600mm	nr	1.00	7.00	178.54	27.83	213.37
900 x 600mm	nr	1.16	8.12	233.28	36.21	277.61
900 x 900mm	nr	1.33	9.31	275.57	42.73	327.61
1200 x 900mm	nr	1.50	10.50	296.94	46.12	353.56
1200 x 1200mm	nr	1.66	11.62	310.93	48.38	370.93
1800 x 1200mm	nr	2.00	14.00	419.79	65.07	498.86
Single skin clear or diffused acrylic pyramidal rooflight						
600 x 600mm	nr	1.00	7.00	182.21	28.38	217.59
900 x 600mm	nr	1.16	8.12	239.68	37.17	284.97
900 x 900mm	nr	1.33	9.31	282.68	43.80	335.79
1200 x 900mm	nr	1.50	10.50	256.85	40.10	307.45
1200 x 1200mm	nr	1.66	11.62	321.48	49.97	383.07
1800 x 1200mm	nr	2.00	14.00	436.58	67.59	518.17

Note: Add 20% to basic material costs for single skin tinted acrylic domed or pyramidal rooflight

Double skin domed rooflight with clear PVC-U standard inner skin and clear, diffused or tinted PVC-U outer skin						
600 x 600mm	nr	1.00	7.00	203.57	31.59	242.16
900 x 600mm	nr	1.16	8.12	279.45	43.14	330.71
900 x 900mm	nr	1.33	9.31	322.77	49.81	381.89
1200 x 900mm	nr	1.50	10.50	343.81	53.15	407.46

ROOFLIGHTS

	Unit	Labour hours	Net labour (£)	Net material (£)	O'heads /profit (£)	Total (£)
1200 x 1200mm	nr	1.66	11.62	365.87	56.62	434.11
1800 x 1200mm	nr	2.00	14.00	538.53	82.88	635.41

Double skin pyramidal rooflight with clear PVC-U standard inner skin and clear, diffused or tinted PVC-U outer skin

600 x 600mm	nr	1.00	7.00	212.46	32.92	252.38
900 x 600mm	nr	1.16	8.12	296.89	45.75	350.76
900 x 900mm	nr	1.33	9.31	340.58	52.48	402.37
1200 x 900mm	nr	1.50	10.50	363.88	56.16	430.54
1200 x 1200mm	nr	1.66	11.62	387.66	59.89	459.17
1800 x 1200mm	nr	2.00	14.00	579.18	88.98	682.16

Double skin domed rooflight with clear PVC-U standard inner skin and clear or diffused wire laminate outer skin

600 x 600mm	nr	1.00	7.00	231.28	35.74	274.02
900 x 600mm	nr	1.16	8.12	314.32	48.37	370.81
900 x 900mm	nr	1.33	9.31	388.47	59.67	457.45
1200 x 900mm	nr	1.50	10.50	395.58	60.91	466.99
1200 x 1200mm	nr	1.66	11.62	457.09	70.31	539.02
1800 x 1200mm	nr	2.00	14.00	607.87	93.28	715.15

Double domed rooflight with clear PVC-U standard inner skin and clear or diffused acrylic outer skin

600 x 600mm	nr	1.00	7.00	219.35	33.95	260.30
900 x 600mm	nr	1.16	8.12	273.47	42.24	323.83
900 x 900mm	nr	1.33	9.31	344.35	53.05	406.71
1200 x 900mm	nr	1.50	10.50	359.85	55.55	425.90
1200 x 1200mm	nr	1.66	11.62	416.83	64.27	492.72
1800 x 1200mm	nr	2.00	14.00	566.16	87.02	667.18

RATES FOR MEASURED WORK

'Coxdome Mark 4' (cont'd)	Unit	Labour hours	Net labour (£)	Net material (£)	O'heads /profit (£)	Total (£)

Double skin pyramidal rooflight with clear PVC-U standard inner skin and clear or diffused acrylic outer skin

600 x 600mm	nr	1.00	7.00	212.72	32.96	252.68
900 x 600mm	nr	1.16	8.12	296.03	45.62	349.77
900 x 900mm	nr	1.33	9.31	343.86	52.98	406.15
1200 x 900mm	nr	1.50	10.50	367.87	56.76	435.13
1200 x 1200mm	nr	1.66	11.62	395.20	61.02	467.84
1800 x 1200mm	nr	0.00	0.00	583.38	87.51	670.89

Note: Add 20% to basic material costs for tinted acrylic outer skin

'Coxdome 2000' rooflight fixed to PVC-U adaptor plugged and screwed to flat builder's curb as distributed by Coxdome Ltd

Double skin domed rooflight with clear PVC-U inner skin and clear, diffused or tinted PVC-U outer skin

600 x 600mm	nr	1.00	7.00	164.88	25.78	197.66
900 x 600mm	nr	1.16	8.12	223.10	34.68	265.90
900 x 900mm	nr	1.33	9.31	234.24	36.53	280.08
1200 x 900mm	nr	1.50	10.50	277.78	43.24	331.52
1200 x 1200mm	nr	1.66	11.62	321.64	49.99	383.25
1800 x 1200mm	nr	2.00	14.00	443.62	68.64	526.26

Double skin pyramidal rooflight with clear PVC-U inner skin and clear, diffused or tinted PVC-U outer skin

600 x 600mm	nr	1.00	7.00	182.37	28.41	217.78
900 x 600mm	nr	1.16	8.12	248.78	38.53	295.43
900 x 900mm	nr	1.33	9.31	260.88	40.53	310.72
1200 x 900mm	nr	1.50	10.50	310.82	48.20	369.52
1200 x 1200mm	nr	1.66	11.62	359.47	55.66	426.75

ROOFLIGHTS

	Unit	Labour hours	Net labour (£)	Net material (£)	O'heads /profit (£)	Total (£)
Double skin domed rooflight with clear acrylic inner skin and clear or diffused acrylic outer skin						
600 x 600mm	nr	1.00	7.00	171.07	26.71	204.78
900 x 600mm	nr	1.16	8.12	213.42	33.23	254.77
900 x 900mm	nr	1.33	9.31	248.93	38.74	296.98
1200 x 900mm	nr	1.50	10.50	285.53	44.40	340.43
1200 x 1200mm	nr	1.66	11.62	351.61	54.48	417.71
1800 x 1200mm	nr	2.00	14.00	495.18	76.38	585.56
Double skin pyramidal rooflight with clear acrylic inner skin and clear or diffused acrylic outer skin						
600 x 600mm	nr	1.00	7.00	189.58	29.49	226.07
900 x 600mm	nr	1.16	8.12	237.64	36.86	282.62
900 x 900mm	nr	1.33	9.31	277.78	43.06	330.15
1200 x 900mm	nr	1.50	10.50	319.70	49.53	379.73
1200 x 1200mm	nr	1.66	11.62	394.23	60.88	466.73
Note: Add 20% to basic material costs for tinted acrylic outer skin						
Double skin domed rooflight with clear wired laminate PVC-U inner skin and clear or diffused PVC-U outer skin						
600 x 600mm	nr	1.00	7.00	180.38	28.11	215.49
900 x 600mm	nr	1.16	8.12	278.21	42.95	329.28
900 x 900mm	nr	1.33	9.31	288.59	44.68	342.58
1200 x 900mm	nr	1.50	10.50	315.45	48.89	374.84
1200 x 1200mm	nr	1.66	11.62	371.95	57.54	441.11
1800 x 1200mm	nr	2.00	14.00	528.33	81.35	623.68

RATES FOR MEASURED WORK

'Coxdome Mark 2000' (cont'd)	Unit	Labour hours	Net labour (£)	Net material (£)	O'heads /profit (£)	Total (£)
Double skin domed rooflight with clear PVC-U inner skin and clear or diffused polycarbonate outer skin						
600 x 600mm	nr	1.00	7.00	231.07	35.71	273.78
900 x 600mm	nr	1.16	8.12	368.67	56.52	433.31
900 x 900mm	nr	1.33	9.31	381.42	58.61	449.34
1200 x 900mm	nr	1.50	10.50	450.41	69.14	530.05
1200 x 1200mm	nr	1.66	11.62	535.97	82.14	629.73
1800 x 1200mm	nr	2.00	14.00	771.94	117.89	903.83

'Coxdome 2000' rooflight clipped to external PVC-U splayed upstand plugged and screwed to roof deck as distributed by Coxdome Ltd

Double skin domed rooflight with clear PVC-U inner skin and clear, diffused or tinted PVC-U outer skin

600 x 600mm	nr	1.00	7.00	237.53	36.68	281.21
900 x 600mm	nr	1.16	8.12	305.71	47.07	360.90
900 x 900mm	nr	1.33	9.31	329.33	50.80	389.44
1200 x 900mm	nr	1.50	10.50	378.84	58.40	447.74
1200 x 1200mm	nr	1.66	11.62	328.52	51.02	391.16
1800 x 1200mm	nr	2.00	14.00	579.18	88.98	682.16

Double skin pyramidal rooflight with clear PVC-U inner skin and clear, diffused or tinted PVC-U outer skin

600 x 600mm	nr	1.00	7.00	253.94	39.14	300.08
900 x 600mm	nr	1.16	8.12	331.38	50.92	390.42
900 x 900mm	nr	1.33	9.31	357.04	54.95	421.30
1200 x 900mm	nr	1.50	10.50	411.88	63.36	485.74
1200 x 1200mm	nr	1.66	11.62	474.25	72.88	558.75

ROOFLIGHTS

	Unit	Labour hours	Net labour (£)	Net material (£)	O'heads /profit (£)	Total (£)

Double skin domed rooflight with clear acrylic inner skin and clear or diffused acrylic outer skin

600 x 600mm	nr	1.00	7.00	242.64	37.45	287.09
900 x 600mm	nr	1.16	8.12	296.03	45.62	349.77
900 x 900mm	nr	1.33	9.31	344.02	53.00	406.33
1200 x 900mm	nr	1.50	10.50	386.59	59.56	456.65
1200 x 1200mm	nr	1.66	11.62	470.33	72.29	554.24
1800 x 1200mm	nr	2.00	14.00	630.74	96.71	741.45

Double skin pyramidal rooflight with clear acrylic inner skin and clear or diffused acrylic outer skin

600 x 600mm	nr	1.00	7.00	261.16	40.22	308.38
900 x 600mm	nr	1.16	8.12	320.24	49.25	377.61
900 x 900mm	nr	1.33	9.31	372.87	57.33	439.51
1200 x 900mm	nr	1.50	10.50	420.76	64.69	495.95
1200 x 1200mm	nr	1.66	11.62	529.79	81.21	622.62

Note: Add 20% to basic material costs for tinted acrylic outer skin

Double skin domed rooflight with clear wired laminate PVC-U inner skin and clear or diffused PVC-U outer skin

600 x 600mm	nr	1.00	7.00	251.95	38.84	297.79
900 x 600mm	nr	1.16	8.12	360.81	55.34	424.27
900 x 900mm	nr	1.33	9.31	383.68	58.95	451.94
1200 x 900mm	nr	1.50	10.50	416.51	64.05	491.06
1200 x 1200mm	nr	1.66	11.62	486.47	74.71	572.80
1800 x 1200mm	nr	2.00	14.00	663.88	101.68	779.56

RATES FOR MEASURED WORK

'Coxdome 2000' (cont'd)	Unit	Labour hours	Net labour (£)	Net material (£)	O'heads /profit (£)	Total (£)

Double skin domed rooflight with clear PVC-U inner skin and clear or diffused polycarbonate outer skin

600 x 600mm	nr	1.00	7.00	302.64	46.45	356.09
900 x 600mm	nr	1.16	8.12	451.27	68.91	528.30
900 x 900mm	nr	1.33	9.31	476.51	72.87	558.69
1200 x 900mm	nr	1.50	10.50	551.47	84.30	646.27
1200 x 1200mm	nr	1.66	11.62	650.49	99.32	761.43
1800 x 1200mm	nr	2.00	14.00	907.49	138.22	1059.71

'VELUX' ROOF WINDOWS

Other types of glass available include: Georgian wired cast and polished; toughened and laminated and obscure

'Velux' roof windows type GGL pivoted; fixing brackets screwed to sloping rafters; aluminium clad externally

Double glazed 2 x 3mm panels clear float glass

GGL1; 780 x 980mm	nr	4.00	28.00	138.18	24.93	191.11
GGL2; 780 x 1400mm	nr	4.50	31.50	166.80	29.75	228.05
GGL3; 940 x 1600mm	nr	5.00	35.00	198.88	35.08	268.96
GGL4; 1140 x 1180mm	nr	4.75	33.25	188.77	33.30	255.32
GGL5; 700 x 1180mm	nr	4.25	29.75	145.34	26.26	201.35
GGL6; 550 x 980mm	nr	3.50	24.50	120.41	21.74	166.65
GGL7; 1340 x 980mm	nr	4.75	33.25	191.24	33.67	258.16
GGL8; 1340 x 1400mm	nr	5.50	38.50	225.78	39.64	303.92
GGL9; 550 x 700mm	nr	3.50	24.50	147.12	25.74	197.36

Double glazed; anti-sun glass and 3mm clear float glass

GGL1; 780 x 980mm	nr	4.00	28.00	162.09	28.51	218.60
GGL2; 780 x 1400mm	nr	4.50	31.50	195.13	33.99	260.62
GGL3; 940 x 1600mm	nr	5.00	35.00	236.89	40.78	312.67
GGL4; 1140 x 1180mm	nr	4.75	33.25	223.68	38.54	295.47

ROOFLIGHTS

	Unit	Labour hours	Net labour (£)	Net material (£)	O'heads /profit (£)	Total (£)
GGL5; 700 x 1180mm	nr	4.25	29.75	169.58	29.90	229.23
GGL6; 550 x 980mm	nr	3.50	24.50	140.86	24.80	190.16
GGL7; 1340 x 980mm	nr	4.75	33.25	225.33	38.79	297.37
GGL8; 1340 x 1400mm	nr	5.50	38.50	269.45	46.19	354.14
GGL9; 550 x 700mm	nr	3.50	24.50	123.17	22.15	169.82

'Velux' roof windows type GHL, top lining fixing brackets screwed to sloping rafters; aluminium clad externally

Double glazed; 2 x 3mm clear float glass

GHL1; 780 x 980mm	nr	4.25	29.75	120.54	22.54	172.83
GHL2; 780 x 1400mm	nr	4.50	31.50	145.51	26.55	203.56
GHL4; 1140 x 1180mm	nr	4.75	33.25	173.49	31.01	237.75
GHL5; 700 x 1180mm	nr	4.25	29.75	173.49	30.49	233.73
GHL7; 1340 x 980mm	nr	4.75	33.25	164.66	29.69	227.60
GHL8; 1340 x 1400mm	nr	5.50	38.50	105.04	21.53	165.07

Double glazed; anti-sun glass and 3mm clear float glass

GHL1; 780 x 980mm	nr	4.00	28.00	181.30	31.39	240.69
GHL2; 780 x 1400mm	nr	4.50	31.50	212.22	36.56	280.28
GHL4; 1140 x 1180mm	nr	4.75	33.25	238.47	40.76	312.48
GHL5; 700 x 1180mm	nr	4.25	29.75	189.98	32.96	252.69
GHL7; 1340 x 980mm	nr	4.75	33.25	238.41	40.75	312.41
GHL8; 1340 x 1400mm	nr	5.50	38.50	280.43	47.84	366.77

Emergency Exit windows also available in this type

'Velux' roof windows type VFE vertical window element; fixed to brickwork reveals

Aluminium clad externally

VFE1/2 60; 780 x 600mm	nr	3.75	26.25	109.34	20.34	155.93
VFE1/2 95; 780 x 950mm	nr	4.00	28.00	122.51	22.58	173.09
VFE3 60; 940 x 600mm	nr	4.10	28.70	109.34	20.71	158.75
VFE3 95; 940 x 800mm	nr	4.25	29.75	122.51	22.84	175.10
VFE4 60; 1140 x 600mm	nr	4.75	33.25	121.49	23.21	177.95

RATES FOR MEASURED WORK

'Velux' roof windows (cont'd)	Unit	Labour hours	Net labour (£)	Net material (£)	O'heads /profit (£)	Total (£)
VFE4 95; 1140 x 800mm	nr	5.00	35.00	135.34	25.55	195.89
VFE7/8 60; 1340 x 600mm	nr	4.75	33.25	134.48	25.16	192.89
VFE7/8 95; 1340 x 800mm	nr	5.00	35.00	150.07	27.76	212.83

Velux aluminium preformed flashings and linings

Type U for tiles and pantiles to suit window size

780 x 980mm	nr	2.00	14.00	25.91	5.99	45.90
780 x 1400mm	nr	2.25	15.75	28.67	6.66	51.08
940 x 1600mm	nr	2.50	17.50	32.10	7.44	57.04
1140 x 1180mm	nr	2.35	16.45	31.41	7.18	55.04
700 x 1180mm	nr	2.10	14.70	25.91	6.09	46.70
550 x 980mm	nr	1.75	12.25	22.83	5.26	40.34
1340 x 980mm	nr	2.35	16.45	32.77	7.38	56.60
1340 x 1400mm	nr	2.75	19.25	35.35	8.19	62.79
550 x 700mm	nr	1.50	10.50	21.11	4.74	36.35

Type H for profiled roofing to suit window size

780 x 980mm	nr	2.00	14.00	29.94	6.59	50.53
780 x 1400mm	nr	2.25	15.75	32.90	7.30	55.95
940 x 1600mm	nr	2.50	17.50	37.79	8.29	63.58
1140 x 1180mm	nr	2.35	16.45	31.00	7.12	54.57
700 x 1180mm	nr	2.10	14.70	26.74	6.22	47.66
550 x 980mm	nr	1.75	12.25	39.06	7.70	59.01
1340 x 980mm	nr	2.35	16.45	39.06	8.33	63.84
1340 x 1400mm	nr	2.75	19.25	41.82	9.16	70.23
550 x 700mm	nr	1.50	10.50	25.04	5.33	40.87

Type L for thin slates to suit window size

780 x 980mm	nr	2.00	14.00	22.75	5.51	42.26
780 x 1400mm	nr	2.25	15.75	26.06	6.27	48.08
940 x 1600mm	nr	2.50	17.50	29.36	7.03	53.89
1140 x 1180mm	nr	2.35	16.45	27.42	6.58	50.45
700 x 1180mm	nr	2.10	14.70	23.92	5.79	44.41
550 x 980mm	nr	1.75	12.25	21.00	4.99	38.24
1340 x 980mm	nr	2.35	16.45	27.60	6.61	50.66
1340 x 1400mm	nr	2.75	19.25	30.71	7.49	57.45

ROOFLIGHTS

	Unit	Labour hours	Net labour (£)	Net material (£)	O'heads /profit (£)	Total (£)
550 x 700mm	nr	1.50	10.50	17.69	4.23	32.42

Type HF for combined GGL and VFE vertical window element 600 or 800mm high

	Unit	Labour hours	Net labour (£)	Net material (£)	O'heads /profit (£)	Total (£)
780 x 980mm	nr	2.50	17.50	35.63	7.97	61.10
780 x 1400mm	nr	2.75	19.25	38.17	8.61	66.03
940 x 1600mm	nr	3.00	21.00	41.04	9.31	71.35
1140 x 1180mm	nr	2.85	19.95	40.37	9.05	69.37
1340 x 980mm	nr	2.85	19.95	40.71	9.10	69.76
1340 x 1400mm	nr	3.25	22.75	43.08	9.87	75.70

Type LF for combined GGL and VFE vertical window element 600 or 800mm high

	Unit	Labour hours	Net labour (£)	Net material (£)	O'heads /profit (£)	Total (£)
780 x 980mm	nr	2.50	17.50	46.98	9.67	74.15
780 x 1400mm	nr	2.75	19.25	46.98	9.93	76.16
940 x 1600mm	nr	3.00	21.00	50.21	10.68	81.89
1140 x 1180mm	nr	2.85	19.95	50.71	10.60	81.26
1340 x 980mm	nr	2.85	19.95	53.31	10.99	84.25
1340 x 1400mm	nr	3.25	22.75	53.09	11.38	87.22

Type HBH 100 for coupled windows horizontally with 100mm frame gap; for profiled roofing materials

	Unit	Labour hours	Net labour (£)	Net material (£)	O'heads /profit (£)	Total (£)
780 x 980mm	nr	2.50	17.50	63.82	12.20	93.52
780 x 1400mm	nr	2.75	19.25	67.69	13.04	99.98
940 x 1600mm	nr	3.00	21.00	74.96	14.39	110.35
1140 x 1180mm	nr	2.85	19.95	74.91	14.23	109.09
700 x 1180mm	nr	2.60	18.20	64.35	12.38	94.93
550 x 980mm	nr	2.25	15.75	57.36	10.97	84.08
1340 x 980mm	nr	2.85	19.95	80.18	15.02	115.15
1340 x 1400mm	nr	3.25	22.75	83.95	16.00	122.70
550 x 700mm	nr	2.00	14.00	54.83	10.32	79.15

RATES FOR MEASURED WORK

	Unit	Labour hours	Net labour (£)	Net material (£)	O'heads /profit (£)	Total (£)
'Velux' interior lining systems						
Type LPH E 30/60 or E 40/50						
780 x 980mm	nr	0.50	3.50	47.36	7.63	58.49
780 x 1400mm	nr	0.75	5.25	52.25	8.62	66.12
940 x 1600mm	nr	1.00	7.00	56.15	9.47	72.62
1140 x 1180mm	nr	1.00	7.00	54.99	9.30	71.29
700 x 1180mm	nr	0.75	5.25	48.72	8.10	62.07
550 x 980mm	nr	0.50	3.50	44.23	7.16	54.89
1340 x 980mm	nr	1.00	7.00	44.41	7.71	59.12
1340 x 1400mm	nr	1.00	7.00	59.68	10.00	76.68
550 x 700mm	nr	0.50	3.50	41.29	6.72	51.51
'Velux' blinds and awnings						
Roller blinds self-coloured for windows size						
780 x 980mm	nr	0.50	3.50	27.43	4.64	35.57
780 x 1400mm	nr	0.50	3.50	33.30	5.52	42.32
940 x 1600mm	nr	0.50	3.50	40.75	6.64	50.89
1140 x 1180mm	nr	0.50	3.50	39.57	6.46	49.53
700 x 1180mm	nr	0.50	3.50	29.38	4.93	37.81
550 x 980mm	nr	0.50	3.50	22.73	3.93	30.16
1340 x 980mm	nr	0.75	5.25	40.36	6.84	52.45
1340 x 1400mm	nr	0.75	5.25	48.20	8.02	61.47
550 x 700mm	nr	0.50	3.50	19.58	3.46	26.54
Venetian blinds standard cords for windows size						
780 x 980mm	nr	1.00	7.00	63.36	10.55	80.91
780 x 1400mm	nr	1.00	7.00	76.04	12.46	95.50
940 x 1600mm	nr	1.00	7.00	91.23	14.73	112.96
1140 x 1180mm	nr	1.00	7.00	85.81	13.92	106.73
700 x 1180mm	nr	1.00	7.00	64.82	10.77	82.59
550 x 980mm	nr	1.00	7.00	56.48	9.52	73.00
1340 x 980mm	nr	1.25	8.75	94.86	15.54	119.15
1340 x 1400mm	nr	1.25	8.75	114.78	18.53	142.06
550 x 700mm	nr	1.00	7.00	49.23	8.43	64.66

ROOFLIGHTS

	Unit	Labour hours	Net labour (£)	Net material (£)	O'heads /profit (£)	Total (£)
Foil coated fabric Siesta blinds for windows size						
780 x 980mm	nr	0.75	5.25	55.94	9.18	70.37
780 x 1400mm	nr	0.75	5.25	70.53	11.37	87.15
940 x 1600mm	nr	0.75	5.25	88.98	14.13	108.36
1140 x 1180mm	nr	0.75	5.25	81.84	13.06	100.15
700 x 1180mm	nr	0.75	5.25	59.82	9.76	74.83
550 x 980mm	nr	0.75	5.25	45.82	7.66	58.73
1340 x 980mm	nr	1.00	7.00	83.02	13.50	103.52
1340 x 1400mm	nr	1.00	7.00	101.48	16.27	124.75
550 x 700mm	nr	0.75	5.25	38.68	6.59	50.52

APPROXIMATE ESTIMATING

Approximate Estimating

Repairs and alterations	Unit	Approximate prices (£)
Take up tile/slate roof coverings from pitched roof	m2	11.00
Take up bituminous felt roof coverings from flat roof	m2	3.50
Take up woodwool slab roof coverings from flat roof	m2	6.50
Take up tile/slate roof coverings from pitched roof; carefully lay aside for re-use	m2	14.00
Take up woodwool slab roof coverings from flat roof; carefully lay aside for re-use	m2	9.00
Inspect roof battens, refix loose and replace with 38 x 25mm		
25% of area	m2	3.00
50% of area	m2	4.00
75% of area	m2	6.00
100% of area	m2	8.00
Remove single broken slate; replace with new Welsh slate size 460 x 230mm	nr	17.00

APPROXIMATE ESTIMATING

Repairs (cont'd)

	Unit	Approximate prices (£)
Remove slates in areas not exceeding 1m2; replace with new Welsh slates size 405 x 205mm	nr	100.00
Remove double course at eaves and refix with new Welsh slates	m	17.00
Remove double course at verge and refix with new Welsh slates	m	22.00
Remove single broken tile; replace with new Marley Modern tile	nr	5.00
Remove tiles in areas not exceeding 1m2; replace with new Marley Modern tiles	nr	22.00
Remove double course at eaves and refix with new Marley Plain tiles	m	9.00
Remove double course at verge and refix with new Marley Plain tiles including bedding and pointing in cement mortar (1:3)	m	14.00
Cut out defective layer of roofing felt in areas not exceeding 1m2; prepare and rebond new layer in hot bitumen	nr	8.00

Slate roofing

	Unit	Approximate prices (£)
Blue/grey slates size 510 x 305mm; 75mm lap, 50 x 25mm softwood battens	m2	65.00

APPROXIMATE ESTIMATING

	Unit	Approximate prices (£)
Slate roofing (cont'd)		
Blue/grey slates size 405 x 205mm; 75mm lap, 50 x 25mm softwood battens	m2	65.00
Asbestos-free cement slates size 600 x 600mm, lap 100mm, 38 x 25mm treated softwood battens	m2	34.00
Asbestos-free cement slates size 400 x 240mm, lap 100mm, 38 x 25mm treated softwood battens	m2	30.00
Tile roofing		
Marley Plain granule or smooth finish tiles size 267 x 165mm; 65mm lap, battens size 38 x 19mm, underfelt	m2	45.00
Marley Mendip tiles size 420 x 330mm; 75mm lap, battens size 38 x 25mm, underfelt	m2	22.00
Redland Plain granular faced or through coloured tiles size 265 x 165mm; 65mm lap, battens size 32 x 19mm, underfelt	m2	45.00
Decking		
'Woodcemair' unreinforced wood-wood-wool slabs (type 5B) in standard lengths, fixed to timber joists, thickness 50mm, (type 500)	m2	16.00

APPROXIMATE ESTIMATING

	Unit	Approximate prices (£)
'Woodcelip' reinforced woodwool slabs, in standard lengths, fixed to timber joists, thickness 50mm (type 503)	m2	28.00

Treated softwood counter battens nailed with galvanized nails to softwood joists size 38 x 19mm, 450mm centres

32 x 19mm	m2	1.75
32 x 25mm	m2	2.00
38 x 22mm	m2	2.10
38 x 25mm	m2	2.40
50 x 25mm	m2	3.00

Sheet metal roofing

Code 4 lead 1.8mm thick, 20.41kg/m2 colour coded blue, fixed with brass screws and copper nails, flat roof with falls not exceeding 10 degrees	m2	85.00
Code 5 lead 2.24mm thick, 25.4kg/m2, colour coded red, fixed with brass screws and copper nails, flat roof with falls not exceeding 10 degrees	m2	95.00
Code 6 lead 2.65mm thick, 30.1kg/m2, colour coded black, fixed with brass screws and copper nails, flat roof with falls not exceeding 10 degrees	m2	105.00
0.55mm thick copper sheeting to BS2870, roof covering, flat	m2	70.00

APPROXIMATE ESTIMATING

Sheet metal roofing (cont'd)	Unit	Approximate prices (£)
0.7mm thick copper sheeting to BS2870, roof covering, flat	m2	80.00
12 gauge zinc (0.65mm thick) to BS849, roof covering, flat	m2	55.00
0.6 commercial grade aluminium to BS1470, roof covering, flat	m2	55.00
Nuralite standard roof sheeting to flat roof	m2	25.00

Fibre cement cladding

Corrugated reinforced cement sheeting, lapped one corrugation at sides and 150mm at ends, fixed with screws and washers to timber purlins

profile 3 grey sheets	m2	20.00
profile 6 grey sheets	m2	18.00

Corrugated glass fibre reinforced translucent sheeting lapped one corrugation at sides and 150mm at ends fixed with screws and washers to timber purlins

profile 3 sheets	m2	26.00
profile 6 sheets	m2	38.00

APPROXIMATE ESTIMATING

	Unit	Approximate prices (£)
Galvanized steel strip troughed sheets; Precision Metal Forming Ltd; Colorcoat HP 200 Leathergrain embossed PVC Plastisol one side, standard light grey backing coat, roof cladding; sloping not exceeding 50 degrees; fixing to purlins with self-tapping screws, 0.7mm 13.5/3 with 19mm corrugation		
roof cladding	m2	15.00

Built up roofing

Built-up bituminous felt roof coverings to areas over 300mm wide laid to falls and crossfalls not exceeding 10 degrees from horizontal

Fibre based granule surfaced felt type 1B weighing 25kg/10m2		
one layer	m2	5.00
two layers	m2	8.00
three layers	m2	12.00
Glass fibre based sand surfaced felt type 3B weighing 18kg/10m2		
one layer	m2	6.00
two layers	m2	10.00
three layers	m2	13.00

APPROXIMATE ESTIMATING

Built-up roofing (cont'd) | Unit | Approximate prices (£)

Polyester based mineral surfaced felt type 5E weighing 38kg/10m2

one layer	m2	11.00

Polyester based 180g/m2 sand surfaced elastomeric bitumen coated felt weighing 40kg/10m2

one layer	m2	8.00

'Foamglas' slabs T2 600 x 450mm laid in hot bitumen on concrete roof

80mm thick	m2	27.00

Rooflights

'Coxdome Mark 1' rooflight

    single skin clear

900mm diameter	nr	80.00

    double skin clear

900mm diameter	nr	165.00

'Coxdome Mark 4' rooflight

    double skin clear

900 x 900mm	nr	360.00

APPROXIMATE ESTIMATING

Rooflights (cont'd)	Unit	Approximate prices (£)
'Velux' roof window double glazed; clear glass		
GHL1; 780 x 980mm	nr	225.00
GHL2; 780 x 1180mm	nr	250.00
GHL4; 1140 x 1180mm	nr	300.00
GHL5; 700 x 1180mm	nr	150.00

Chapter 6
Plant and Tool Hire

The prices contained in this chapter are based upon information supplied by HSS Hire Shops Ltd, 25 Willow Lane, Mitcham, Surrey, who have over 200 shops throughout the country (see Yellow Pages). The prices exclude VAT and delivery charges.

	First 24 hrs (£)	Addit 24 hrs (£)	Per week (£)
ACCESS AND SUPPORT			
Narrow tower base size 1.3 x 1.8m, height			
2.3m	29.00	14.50	58.00
4.3m	42.50	21.25	85.00
6.3m	56.00	28.00	112.00
8.3m	69.50	34.75	139.00
10.3m	83.00	41.50	166.00
Span tower base size 1.5 x 1.8m or 1.5 x 2.5m, height			
2.3m	29.00	14.50	58.00
4.3m	42.50	21.25	85.00
6.3m	56.00	28.00	112.00
8.3m	69.50	34.75	139.00
10.3m	83.00	41.50	166.00

PLANT AND TOOL HIRE

Access and support (cont'd)

	First 24 hrs (£)	Addit 24 hrs (£)	Per week (£)
Push-up ladders			
Double 3.5m extending to 6.2m	9.50	4.75	19.00
5.0m 9.0m	13.00	6.50	26.00
Treble 2.5m 6.0m	9.50	4.75	19.00
3.5m 9.1m	13.00	6.50	26.00
Roof ladders			
Alloy 4.9m, 5.9m and 6.9m	15.60	5.20	26.00
Ladder stay each	5.40	1.80	9.00
Stand off	5.00	2.50	10.00
Wheeled	7.50	3.75	15.00
Builder's steps			
8 Tread height 1.5m	7.50	3.75	15.00
10 Tread height 2.1m	8.50	4.25	17.00
12 Tread height 2.7m	10.00	5.00	20.00
Folding indoor scaffold	20.00	10.00	40.00
Lightweight staging			
Length: 2.4m	8.50	4.25	17.00
3.0m	9.50	4.75	19.00
3.6m	10.50	5.25	21.00
4.2m	12.00	6.00	24.00
4.8m	14.00	7.00	28.00
6.0m	16.00	8.00	32.00
7.2m	22.00	11.00	44.00

PLANT AND TOOL HIRE

	First 24 hrs (£)	Addit 24 hrs (£)	Per week (£)
SAWING AND CUTTING			
Tile saw	36.00	12.00	60.00
Tile cutter			
manual	9.00	3.00	15.00
cordless	19.20	6.40	32.00
breaker	3.00	1.00	5.00
MECHANICAL HANDLING			
Lifter stacker			
SWL 363kg	48.00	16.00	80.00
SWL 295kg	51.60	17.20	86.00
Tile hoist	76.80	25.60	128.00

Chapter 7
General Data

The metric system

Linear

1 centimetre (cm)	= 10 millimetres (mm)
1 decimetre (dm)	= 10 centimetres (cm)
1 metre (m)	= 10 decimetres (dm)
1 kilometre (km)	= 1000 metres (m)

Area

100 sq millimetres	= 1 sq centimetre
100 sq centimetres	= 1 sq decimetre
100 sq decimetres	= 1 sq metre
1000 sq metres	= 1 hectare

Capacity

1 millilitre (ml)	= 1 cubic centimetre (cm^3)
1 centilitre (cl)	= 10 millilitres (ml)
1 decilitre (dl)	= 10 centilitres (cl)
1 litre (l)	= 10 decilitres (dl)

Weight

1 centigram (cg)	= 10 milligrams (mg)
1 decigram (dg)	= 10 centigrams (cg)
1 gram (g)	= 10 decigrams (dg)
1 decagram (dag)	= 10 grams (g)
1 hectogram (hg)	= 10 decagrams (dag)
1 kilogram (kg)	= 10 hectograms (hg)

GENERAL DATA

Imperial/metric conversions

Linear

    1 in = 25.4mm   1mm = 0.03937 in
    1 ft = 304.8mm   1cm = 0.3937 in
    1 yd = 914.4mm   1dm = 3.397 in
                                        1m = 39.37 in

Square

    1 sq in = 645.16mm^2 1cm^2 = 0.155 sq in
    1 sq ft = 0.0929m^2 1m^2 = 10.7639 sq ft
    1 sq yd = 0.8361m^2 1m^2 = 1.196 sq yd

Cube

    1 cu in = 16.3871cm^3 1cm^3 = 0.061 cu in
    1 cu ft = 0.0283m^3 1m^3 = 35.3148 cu ft
    1 cu yd = 0.7646m^3 1m^3 = 1.307954 cu yd

Capacity

    1 fl oz = 28.4ml   1ml = 0.0352 fl oz
    1 pt = 0.568 l   1dl = 3.52 fl oz
    1 gallon = 4.546 l   l litre = 1.7598 pt

Weight

    1 oz = 28.35g   1g = 0.035 oz
    1 lb = 0.4536kg   1kg = 35.274 oz
    1 st = 6.35kg   1t = 2204.6 lb
    1 ton = 1.016t   1t = 0.9842 ton

Temperature equivalents

In order to convert Fahrenheit to Celsius deduct 32 and multiply by 5/9. To convert Celsius to Fahrenheit multiply by 9/5 and add 32.

GENERAL DATA

Temperature equivalents (cont'd)

Fahrenheit	Celsius
230	110.0
220	104.4
210	98.9
200	93.3
190	87.8
180	82.2
170	76.7
160	71.1
150	65.6
140	60.0
130	54.4
120	48.9
110	43.3
100	37.8
90	32.2
80	26.7
70	21.1
60	15.6
50	10.0
40	4.4
30	-1.1
20	-6.7
10	-12.2
0	-17.8

Number of slates per m2

Asbestos-free

Size (mm)	Lap (mm)	Nr of slates
400 x 200	70	30.0
400 x 200	76	30.9
400 x 200	90	32.3
400 x 240	80	26.1
500 x 250	90	19.5
500 x 250	80	19.1
500 x 250	70	18.6
500 x 250	76	18.9

GENERAL DATA

Number of slates per m2 (cont'd)

Asbestos-free

500 x 250	90	19.5
500 x 250	106	20.5
500 x 250	100	20.0
600 x 300	106	13.6
600 x 300	100	13.4
600 x 300	90	13.1
600 x 300	80	12.9
600 x 300	70	12.7
600 x 350	100	11.5

Blue Welsh

Size (mm)	Nr of slates
405 x 205 (16" x 8")	29.59
405 x 255 (16" x 10")	23.75
405 x 305 (16" x 12")	19.00
460 x 230 (18" x 9")	23.00
460 x 255 (18" x 10")	20.37
460 x 305 (18" x 12")	17.00
510 x 255 (20" x 10")	18.02
510 x 305 (20" x 12")	15.00
560 x 280 (22" x 11")	14.81
560 x 305 (22" x 12")	14.00
610 x 305 (24" x 12")	12.27

Westmoreland Green

1 ton (Imperial) standard quality covers approximately 18 to 20m2

1 ton (Imperial) peggies covers approximately 15 to 16m2

GENERAL DATA

Marley tiles (100mm gauge)

Type	Nr/m2
Plain	60.0
Feature	56.0
Ludlow Plus	17.4
Anglia Plus	17.3
Ludlow Major	10.7
Mendip	10.6
Double Roman	10.4
Modern	10.8
Wessex	11.0
Bold Roll	10.6
Monarch	14.5

Redland tiles (100mm gauge)

Type	Nr/m2
Renown	9.7
49	16.3
50 Double Roman	9.7
Norfolk Pantiles	16.3
Stonewold	8.2
Richmond	11.1
Saxon	11.1
Delta	8.2
Plain	60.0
Ornamental	52.7
Downland	60.0
Cambrian	13.3
Rosemary	60.0

Lead

Code	Nr/m2
3	14.97
4	20.41
5	25.40
6	30.05
7	36.72
8	40.26

Index

Accountant, 6, 10
Advertising, 2
All-in rate, 42-43
Alterations, 257-258
Aluminium
  flashings, 55-57, 126-131
  sheeting, 47-49, 73-74, 112-117, 191-193

Bad debts, 37
Banks, 4, 10
Battens, 70, 165-166
Break-even point, 25-27
Built-up roofing, 74-75, 207-223, 262-263

Copper sheeting, 73, 194-196, 260, 260-261

Discounts, 47
DOE, 2-3
DSS, 5
DTI, 6
Duracem slates, 62, 167-168
                Lorries, 45
Estimating, 35-40
Eternit 2000
  slates, 62, 169-170

Fibre bitumen
  sheeting 71-73, 200-203

Fibre cement
  sheeting, 70-71, 111, 261
  slates, 62-63, 167-170
Finance, 1, 8
Foamglas, 75-76

Galvanized steel
  flashings, 57-61, 131-133
  sheeting, 49-54, 117-126
General data, 269-273
Glass-reinforced cladding, 71, 134, 261
Grants, 11

HSE, 8

Inspector of Taxes, 4
Insulation, 54, 126
Insurance broker, 7-8

Labour, 36
Labour hours, 42-43
Labour rate, 42-43
Lead sheeting, 73, 183-190, 260
Lining panels, 102, 252-253

Market, 1
Marketing, 19-20
Marley
  reconstructed slates, 63, 180
  tiles, 63-66, 153-165, 259

INDEX

Material costs, 47-102
Materials 37, 43
Metal profiled sheeting, 47-54, 112-133
Metric system, 269-270

Natural slates, 62, 173-178, 258-259
Nuralite, 71-73, 200-203, 261

Overheads
  generally, 37, 44
  head office, 38-40
  site, 37-38

Partnerships, 29
Plant, 37, 265-267
Profitability, 21-24
Profits, 40, 44

Redland
  reconstructed slates, 63, 180, 181
  tiles, 66-67, 153-165, 259
Regional variations, 41
Repairs, 257-258
Rivendale slates, 62, 171
Rooflights, 77-97, 227-248, 263
Rubbish, 44-45

Shingles, 67, 182
Skips, 44-45
Slates
  Country, 62, 171-172
  Duracem, 62, 167-168
  Eternit 2000, 62, 169-170
  fibre cement, 62-63, 167-170
  natural, 62, 173-178, 258-259
  reconstructed stone, 62, 180-181
  Rivendale, 62, 171
  Welsh, 62, 173-178

Westmoreland, 62, 178-179
Small Firms Service, 2
SMM6/7 table, vii-viii
Solicitor, 7
Steel flashings, 57-61, 131-133
Steel sheeting, 49-54, 117-126
Suppliers, 3

Tax, contractor's (714), 12-15
Taxation
  generally, 29-37
  allowances, 33
  capital gains, 34-35
  entertainment, 35
  inspector, 4
  joint income, 34
  mortgage relief, 34
  rates, 32
Tiling
  Marley, 63-66, 135-152, 259
  Redland, 66-67, 153-165, 259
Training, 2
Translucent sheeting, 71, 134, 261

Underfelt, 70, 166

VAT, 4-5, 16-18, 37
Velux
  blinds, 102, 252-253
  flashings, 100-101, 250-251
  linings, 102, 252-253
  windows, 98-99, 248-250, 264

Wage awards, 42
Woodcelip decking, 68-70, 105-107, 260
Woodcemair decking, 68, 105, 259
Woodwool decking, 68-70, 105-107, 259

Zinc sheeting, 73, 197-199, 261